100 YEARS OF BUILDING

San Antonio

The People Who Built the Seventh Largest City in the USA, 1923–2023

Praise for *100 Years of Building San Antonio*

"*100 Years of Building San Antonio* is a sweeping chronicle of
San Antonio's modern physical profile with gems of anecdotes and
historical details about the construction companies and the
workers who built the famous buildings, decade by decade."

— *Stephen Amberg, Ph.D.,*
The University of Texas at San Antonio

"Doug and Michele have written a great book about the major building
projects in San Antonio over the last hundred years. Building for the
future drives our local economy and, in this book, they have captured
the major projects that have positioned San Antonio as a fast-growing
economy."

— *Nelson Wolff, Bexar County Judge 2001-2022*

"A city can be defined by its buildings, as well as its people.
This book centers on the brick-and-mortar construction, but also
contains a goodly amount of history of the people who built the
buildings, and the history surrounding why they were built.
It also relates the occasional controversy that surrounds great
projects. It's all mighty interesting. After you have read this book,
you will know many things about San Antonio that you did not know
before. I know I did, and recommend you do the same."

— *Phil Hardberger, Mayor of San Antonio 2005-2009*

"Doug and Michele's narrative is a splendid journey that connects iconic landmarks to the builders whose artistry created our distinct and unique city."

— Richard Perez, President & CEO,
San Antonio Chamber of Commerce

"Doug and Michele McMurry artfully weave the stories of San Antonio's familiar and famous places as they chronicle our city's development over the last century."

— Torrey Stanley Carleton, Hon. AIA,
Executive Director, AIA San Antonio

ISBN: 978-1-958407-03-5 (Hardback)

ISBN: 978-1-958407-04-2 (Soft Cover)

Front cover photos, clockwise from the top: *Guido working crew; aerial view of Trinity University; George W. Mitchell and his wife, Iris; Alamodome; Nunnelly groundbreaking; Credit Human Headquarters; Tower of the Americas under construction; Robert L. B. Tobin Land Bridge; Richard Alterman with his mother.*

Back cover: *F. A. Nunnelly General Contractor pickup with local children.*

Photo Credits: The Associated General Contractors; BakerTriangle; Bartlett Cocke General Contractors; City of San Antonio; The Conservation Society of San Antonio; G. W. Mitchell Construction; Guido Construction; Joeris General Contractors; Nunnelly General Contractor; UTSA/San Antonio Express-News/ZUMA Press.

Book design by designpanache

ELM GROVE PUBLISHING

San Antonio, Texas, USA
www.elmgrovepublishing.com

Elm Grove Publishing is a legally registered trade name of Panache Communication Arts, Inc.

100 YEARS OF BUILDING
San Antonio
The People Who Built the Seventh Largest City in the USA, 1923–2023

Doug McMurry, Hon. AIA San Antonio
Michele McMurry

Contents

For the craftworkers who do the heavy lifting,
building the city one story at a time.

Foreword
By Ron Nirenberg

From the time of settlement by indigenous Native Americans, the place we now call San Antonio was a crossroads—a confluence of cultures in the Americas where land and abundant water brought communities together. As the years unfolded, many people would arrive and leave their mark on the landscape, including the eighteenth-century Spanish colonists who established Mission de Valero, along with four other missions: San José, Concepción, San Juan, and Espada. The chapel of Mission de Valero—better known as the Alamo—is one of the most recognized structures in the world and, with the other San Antonio missions, was designated in 2015 as the first and only UNESCO World Heritage site in the state of Texas.

While this city is recognized internationally because of the architecture that marked its early history, San Antonio did not stop growing, developing, and attracting people from across the globe to visit its unique structures and buildings. Today, San Antonio remains the most visited city in Texas—more than forty million people annually—in part because of unique treasures like the River Walk, itself one of the top sites in the country.

The story of San Antonio's transition from "Cradle of Texas Liberty" to urban metropolis with the nation's seventh largest population can be witnessed through the iconic structures that dot the flat plains of this South Texas basin. After the calamitous flood of 1921 devastated downtown buildings and killed more than fifty people in the city's West Side, a dam was constructed, forming the foundation of a flood control system that precipitated the construction of the River Walk, the Tower of the Americas, and countless other icons of the 1968 World's Fair era.

Today, San Antonio is still a city best known for the Alamo, but a closer look reveals a community as unique in its modernity as in its history. Military City USA® in 2023 also boasts everything from culinary and renown, diverse public art, to world-leading cyber security, education, and biomedical research facilities. Indeed, the last one hundred years of development in the post-World War I period deserve to be remembered for bringing San Antonio into the future.

100 Years of Building San Antonio captures that remarkable transition by memorializing the people and companies behind it. The journey begins just as the Associated General Contractors (AGC) San Antonio Chapter, which represents more than 400 companies in the nineteen-county South Texas region, was born in 1923. Author Doug McMurry, in his three decades of leading the San Antonio AGC formed many close personal and professional relationships with these builders of our city, and it is through his love of what San Antonio has become that he encourages us to remember their names, just as we recognize their work. Along with co-author Michele McMurry, their account is a compendium of the modern built environment and an homage to the community that gave rise to it.

As many have rightly asked before, "What is a city, if not just a collection of roadways, bridges and buildings?" From icons such as the Tower Life Building and the national historic landmark Jefferson High School to the Alamodome and the Tower of the Americas, this story of San Antonio provides an elegant appreciation of the people and cultures that built one of America's most unique cities.

Ron Nirenberg

Introduction

San Antonio looked quite different one hundred years ago. There were no tall buildings, freeway overpasses, or corporate headquarter campuses. The international airport, South Texas Medical Center, and University of Texas at San Antonio did not exist. Nor was the city home to the largest concentration of military bases in the United States. Imagine a time without H-E-B stores, amusements parks, or shopping centers like North Star Mall.

Despite the lack of public works infrastructure a century ago, San Antonio was at that time the largest city in Texas. Economic optimism and transformation, unfolding following World War I and prior to the onset of the Great Depression, were reflected in new buildings and the people who erected them. Pleased with the city's growth, Mayor Charles McClellan Chambers proclaimed during the 1929 opening of the Smith-Young Tower, "The faster the Rip Van Winkles of this city go, the faster we will progress." More than seventy years later, reporter Benjamin Olivo would reflect on this unprecedented boom in San Antonio's history.

This book is a story about how San Antonio became the nation's seventh-largest city. It is also about how we invested in ourselves and the physical environment that defines us. Ultimately, it is a fascinating account of buildings, determined individuals, and successful construction companies.

Many of these companies are members of the San Antonio Chapter of the Associated General Contractors. The year 2023 marks the chapter's centennial, a remarkable achievement for a trade association dedicated to serving not only general contractors but also the industry's suppliers and subcontractors.

Focusing mainly on commercial construction and noteworthy renovations of iconic buildings, we explore who built the familiar and famous places including revered landmarks, such as Trinity University, the Tower Life Building, and USAA's sprawling headquarters, as well as the Alamodome, SeaWorld San Antonio, and the "enchilada red" Central Library. We also recount how some buildings, like HemisFair Arena and McCreless Shopping City, have been lost to history.

Ultimately, by charting the last hundred years of building San Antonio, we offer a prelude to the future and history yet written.

Chapter One
1923–1932

Familiar & Famous Places	Medical Arts Building (Emily Morgan Hotel)
	San Antonio Builders Exchange
	Olmos Dam
	Milam Building
	Atkinson Residence (McNay Art Museum)
	Smith-Young Tower (Tower Life Building)
	Express Publishing Company (San Antonio Express-News)
	San Pedro Playhouse (The Public Theater of San Antonio)
	Thomas Jefferson High School

Our story begins in 1923 with landmark construction unfolding across the country amid a bustling, post-World War I economy. The iconic Hollywood Sign atop Southern California's Mount Lee reaches its painstaking completion, as does New York's Yankee Stadium as the world's largest baseball park.[1]

In Texas, the San Antonio Chapter of the Associated General Contractors (AGC) organizes one year before the forming of the San Antonio Conservation Society in 1924. Together, the two groups will help build and preserve this city's most iconic structures for the next one hundred years.

Nathan Alterman Electric Company, too, gets its start in 1923 when twenty-four-year-old Nathan Alterman, the son of Russian immigrants who came to San Antonio in the 1880s to run a dry goods store, purchases an existing business.[2] His crew travels to jobs via trolley cars carrying their ladders, tools, and materials. Someday, the small electrical firm will evolve into one of San Antonio's most prominent electrical contractors, employing hundreds and helping to build noteworthy city projects including SeaWorld San Antonio and Toyota Motor Manufacturing, Texas.

❖ ❖ ❖

About this time, Alamo Heights developer Clifton George and Texas architect Ralph H. Cameron conceive the striking, Gothic Revival–style Medical Arts Building. Contractors break ground in 1923 on a triangular site just steps from the Alamo and finish construction in 1926. Structural engineering pioneer W. E. Simpson, a San Antonio native whose résumé includes the St. Anthony Hotel and the relocation of the Alamo National Bank building during the widening of Commerce Street in 1913, provides structural framework to the hexagonal tower that modernizes the city's most historic plaza. Many of San Antonio's physicians will build long careers here before a two-year renovation in 1984 converts it into the Emily Morgan Hotel.[3]

❖ ❖ ❖

Building activity is on the rise in 1925, and local leaders, including Mayor John W. Tobin project that San Antonio is on "the threshold of what gives every promise of being the most prosperous year in its history."[4] The San Antonio Builders Exchange welcomes the completion of its first brick-and-mortar home this year. The ten-story, brick, reinforced concrete and steel building at the corner of St. Mary's and Pecan streets attracts attention beyond San Antonio as the first single structure in the southern region to house the "heart of the industry." The central gathering space includes a library, exhibit hall, plan room, and rooftop garden, and it "bespeaks the spirit of co-operation and mutual service which has developed within the various component trade units constituting the building industry in this city."[5] James Aiken is the general contractor. Subcontractors on the project include Fred Geyer Plumbing, Alamo Brick Company, and Alamo Iron Works.

❖   ❖   ❖

Part of a $5 million bond issue is devoted to flood prevention, and in 1926, Olmos Dam comes to fruition in the aftermath of the catastrophic flood of 1921 that killed dozens of people and cost the city millions in repairs.[6] C. D. Orchard, an engineer and artifact enthusiast, collects and meticulously documents specimens from the Paleoindian period discovered during the project.[7] McKenzie Construction Company serves as general contractor. The retention dam, built to prevent the rapid flow of floodwaters into the San Antonio River, will undergo improvements by the City of San Antonio and the San Antonio River Authority in the 1970s and 1980s. Archer Western Construction will complete additional upgrades to the crown jewel of the city's flood protection system in 2011.

❖   ❖   ❖

The Milam Building, built by L. T. Wright and Company, opens in 1928 as another downtown expression of Roaring Twenties modernity in design and construction sweeping across the United States and Europe.[8] At twenty-one stories, it surpasses the Medical Arts Building as the city's loftiest, stands also as the tallest brick and reinforced-concrete structure in the country and its first air-conditioned skyscraper.[9]

❖ ❖ ❖

During this same year, George W. Mitchell, founder of G. W. Mitchell Construction, builds the lavish Atkinson Residence gracing twenty-three acres at the intersection of North New Braunfels Avenue and Austin Highway. The stucco mansion, adorned with intricate stenciling, represents the finest in Spanish Colonial Revival homes.[10] After Marion Koogler McNay's death in 1950, most of the home's twenty-four rooms will house her vast art collection as the McNay Art Museum, the first museum of modern art in Texas.[11]

❖ ❖ ❖

A year later, San Antonio reaches new heights and acclaim with the thirty-one-story Smith-Young Tower, the tallest building in Texas and even west of the Mississippi River.[12] The landmark downtown project—another built by McKenzie Construction Company and engineered by W. E. Simpson Company—attracts widespread attention and publicity, as reporters buzz about how it marks the transformation of San Antonio into a growing metropolis.[13] Not only statuesque, the high-rise also marks an advancement in construction for its use of under-reamed piers, drilled fifty feet below street level.[14] Pleased with the undertaking and growth of the city, Mayor C. M. Chambers proclaims at the building's opening, "The faster the Rip Van Winkles of this city go, the faster we will progress." Later, on April 7, 2015, reporter Benjamin Olivo will observe, "Throughout the 1920s, San Antonio's skyline saw what at the time was an unprecedented growth spurt: a high-rise boom that yielded such iconic structures as the Smith-Young (Tower Life) and Medical Arts (Emily Morgan Hotel) buildings."[15]

❖ ❖ ❖

A new home for the Express Publishing Company reaches completion just days before the Wall Street Crash of 1929, commanding the corner of Avenue E and Third Street. "The building possesses an architectural beauty that makes it distinctively attractive, and at the same time it stands out as a model of simplicity, in keeping with utilitarian purposes to which it is devoted," according to a history published during the building's dedication.[16] Its builder, Jopling Construction Company, contributes a full-page advertisement in the *San Antonio Express* thanking 300 workers, superintendent of construction A. G. Keller, and project subcontractors, "whose faculty of achieving the best possible results without friction cannot be valued too highly." Jopling prides itself as a member of AGC, citing the organization's tenets of skill, integrity, and responsibility.

❖   ❖   ❖

Employing these tenets, Louis L. Guido, patriarch of the Guido family of general contractors, and partner Vincent Falbo, wrap up construction of the city's first theater, San Pedro Playhouse, at the close of 1929.[17] The columned structure embodies Greek Revival and Neoclassical architecture, lending courtly elegance to San Pedro Springs Park, the state's first park and among the first municipal parks in the country.[18]

❖   ❖   ❖

As part of a $3.7 million school bond proposal, San Antonio Independent School District awards the general construction contract for building Thomas Jefferson High School to Walsh, Burney & Key. No roadways exist beyond Fredericksburg Road, requiring workers to access thirty-three-acre Spanish Acres with mule-drawn rigs to excavate the foundation. Jud & Ormond provides plumbing and heating, and Crowther Electric Company, the electrical needs of the project. After its completion in 1932, "the most outstanding high school in America," according to *Life Magazine*, gains national attention due to a palatial esthetic and elaborate Spanish-Moorish design more characteristic of a

luxury hotel. It will serve as a movie set and the subject of documenta-
ries. In 1983, the San Antonio Historical Society and the San Antonio
City Council will declare the campus a city historic landmark, and the
Texas State Historical Society will follow unanimously with state desig-
nation. The same year, it joins the National Register of Historic Places.[19]

George W. Mitchell and his fellow directors at the San Antonio
Chapter of AGC conclude the organization's first decade with meetings
at the Builders Exchange office, where they plan the future and discuss
how to improve the industry.[20] In the decades that follow, hundreds of
news articles will feature the city's landmark structures. Few will mention who built them.

*Louis L. Guido, patriarch of the Guido family of general contractors,
pictured in his US service uniform*

Nathan Alterman Electric Company was founded in 1923.

*G. W. Mitchell Sr., founder of G. W. Mitchell Construction,
built the Atkinson Residence.*

Chapter Two
1933–1942

Familiar & Famous Places	US Courthouse and Post Office (Hipolito F. Garcia Federal Building and US Courthouse)
	Station Hospital (Brooke Army Medical Center)
	Alamo Stadium
	San Antonio River Walk

The year 1933 marks a turning point in American history with San Antonio a part of shifting fortunes. Franklin D. Roosevelt is sworn in as president and the New Deal, which includes the Civilian Conservation Corps (CCC), Works Progress Administration (WPA), and Public Works Administration (PWA), is about to take off. The New Deal and the flow of federal funding will come to play a significant role in the growth of the city as the new administration seeks to provide relief for the unemployed, recover the national economy, and reform the banking system in the aftermath of the Wall Street Crash of 1929.

Prohibition ends this year and beer sales resume at the San Antonio Brewing Association, an important contributor to San Antonio's economic growth before it becomes more well-known as Pearl Brewing Company. F. A. Nunnelly General Contractor, in years to come, will build most everything at the facility beyond the original brew house and stables. Decades later, Harold "Bubba" Kunz of Kunz Construction Company will admonish his son-in-law, general contractor Mike Cooney, for pursuing subsequent work at the brewery, referring to it as "Nunnelly's work!" This will not be the last time a contractor recognizes territorial boundaries in the local marketplace.

One notable project of the New Deal—the US Courthouse and Post Office—exemplifies the ambition behind federal public works programs implemented to relieve devastation of the Great Depression. The General Services Administration notes, "Its construction accomplished several goals—generating employment, housing all federal agencies in a single building, and streamlining San Antonio's quickly expanding postal needs."[1] Detroit-based builder A. W. Kutsche and Company completes the $2.3 million federal project in 1936, and it is dedicated on March 24, 1937. The six-story and atypical six-sided mega structure— "one of the most beautiful and satisfactory of federal buildings to date," according to government officials—encompasses an entire city block and features steel and concrete clad in Texas pink granite and cream limestone, a polygonal formation, centered on a central light court. Innovative in both design and in use of local materials, it is also marveled as another air-conditioned first in the country.[2]

❖   ❖   ❖

Construction on the 418-bed Station Hospital at Fort Sam Houston begins in 1936 to support the advancement of San Antonio's military medical capabilities. Completed a year later, the $3 million project replaces the original eighty-four-bed Station Hospital from 1908 as "the most modern plant in the army."[3] Banspach Brothers, a local firm, constructs its foundation, and R. E. McKee of El Paso serves as general contractor for the hospital. The replacement hospital will become a central part of Brooke Army Medical Center (BAMC) in 1942 as the renamed Brooke General Hospital, serving as the largest and most significant healthcare organization within the United States Department of

Defense.[4] *San Antonio Express-News* reporter Sig Christenson will reflect years later in a January 31, 2015, story, "How it began and where it would go is one of the city's great stories, for while BAMC was a building, it became much, much more. Over 170 years, San Antonio would become the home of Army medicine, serving as a hub of clinical care where lives are saved and rehabilitated. Over sixty of those years, the ornate seven-story hospital serving as its beating heart."[5]

❖ ❖ ❖

Alamo Stadium, another noteworthy WPA project proposed initially by San Antonio Independent School District trustees in 1939, is erected on the site of an abandoned rock quarry for less than $500,000. The WPA provides most of the funding with approximately $110,000 generated from district revenue bonds. The "rockpile," dedicated on September 20, 1940, to a record gathering of sports fans, will earn addition to the National Register of Historic Places on September 8, 2011. A $35 million renovation in 2013 will restore the facades, plaques, and markers.[6]

❖ ❖ ❖

Long before the San Antonio River Walk becomes the state's top tourist destination, the dark and dangerous place for residents during the 1930s seems more a flooding nuisance. Civic leader A. C. "Jack" White serves as an instrument for change in the passing of a bond issue to support the 1938 San Antonio River beautification project. Continuing its New Deal efforts in the Alamo City, the WPA provides essential funding in 1939 that initiates the construction of stairs, bridges, and walkways, as well as lighting and extensive landscaping including the planting of bald cypress trees that will endure as one of the area's most distinguishing features.[7] Robert H. Turk serves as WPA superintendent, and at its peak the original section of the project—designed by Robert H. H. Hugman—employs 389 workers.[8] As a boy, Richard Alterman remembers the project his family's company, Nathan Alterman Electric Company, worked on and later describes it as "one of the outstanding projects that led to continued advancement to our city." The San Antonio River Walk opens in 1941, the year of the first Fiesta River Parade.

This period ends in 1942 as Nazi Germany, under the command of Adolf Hitler, reaches its peak of dominance in northern Europe during World War II. San Antonio, like the rest of the country, will be forever affected by the war and ensuing geopolitical realignment.

*As a boy, Richard Alterman (pictured with his mother) remembers
his father's company working on the San Antonio River Walk.*

*Old Pearl Brewery entrance. Pearl Brewing Company contributed
to San Antonio's economic growth.*

Floyd Arthur Nunnelly Sr. founded F. A. Nunnelly General Contractor.

Chapter Three
1943–1952

It is January 1943. The headquarters of the United States Department of Defense and the world's largest office building, the Pentagon, is dedicated in Arlington, Virginia. In San Antonio, Kelly Air Force Base plays a significant role in defending the country, training bomber pilots, and becoming a major center for aircraft maintenance as World War II carries on.[1] G. W. Mitchell Construction builds the base's Cylinder Reclamation Building, Armament Repair Building, Kelly Field Water Tank, Air Freight Terminal, and the Sewer Plant Maintenance Building.[2]

The same year, the *San Antonio Express* reports that Lee Christy, a general contractor, is re-elected president of the San Antonio Builders Exchange and G. W. Mitchell as treasurer. Christy predicts the construction industry will be busy in the post-war period.[3] He is right.

Commercial development expands northward in the aftermath of the war, stretching beyond major thoroughfares Broadway Street and San Pedro Avenue, and Interstate Loop 410 before it eventually loops the inner city.[4] But notable projects are cropping up throughout town.

These are the days when H-E-B Grocery Company is building its stores, and the ambitious grocer opens its fourth San Antonio location on April 20, 1945, at 1601 Nogalitos Street on the city's South Side. An advertisement in the *San Antonio Light* billing it as "the store of tomorrow" and promising "extra-large floor space for thousands of food items, spacious vegetable and fruit counters," and "the largest meat refrigeration displays in San Antonio" draws curious customers who crowd the entrance on opening day. The Art Deco-esque addition to one of San Antonio's oldest neighborhoods brings more than ritz and 15,000 square feet of shopping convenience. It unites a community with tradition that will endure through multiple remodels, the first in 1950 that air conditions the space. A 1957 expansion will double its size, and nearly seven decades from now, the Nogalitos location will gain an additional 29,000 square feet on multiple levels while still retaining is historical charm as the city's longest operating H-E-B store.[5]

❖   ❖   ❖

In an underserved area of downtown, G. A. "Tano" Lucchese's Alameda Theater opens on March 10, 1949, with considerable fanfare. The *San Antonio Light* describes the grand building —the largest Spanish-language picture house in the country, seating 2,500 patrons—as having colorful and narrative murals, modern restrooms, an elaborately decorated lobby, and a crying room for children. The "most modern projectors in engineering and design" resemble the sophistication of New York City technology, posing low fire risk. On the exterior, a tricolor, eighty-six-foot-high illuminated sign facing West Houston Street peaks five stories above the theater's roofline as the tallest in San Antonio, surely awing patrons attending the opening gala featuring luminaries from Mexico City. Electrical contractor Shelnutt Neon had installed nearly 4,000 feet of cold cathode fluorescent light tubing to achieve four times the brilliance of that of neon tubes most used before this time.[6] O. L. Edwards served as construction superintendent, and Jno. A. Williamson Company and The Farwell Company helped complete the project. Thirty-three years later, the San Antonio Conservation Society will fight to preserve their good work.

❖ ❖ ❖

Also, in 1949, following the passing of a $1.5 million bond issue, good work begins on the new Alamo Heights High School as architect Bartlett Cocke announces the selection of G. W. Mitchell Construction as general contractor. Cocke explains to the *San Antonio Light* how G. W. Mitchell trimmed its bid by approximately $38,000 on the $737,625 project to remain in consideration. State lawmakers will pass legislation four decades later allowing the awarding of school projects based on best value over lowest price. Bartlett Cocke Jr., a general contractor, will recount in *Bartlett Cocke General Contractors: The First 50 Years, 1959–2009*, "This new law gave us a competitive advantage over some of our competitors."[7] Building plans, according to Alamo Heights Independent School District superintendent Edward T. Robbins, "embody all the best practices in the finest school construction." This includes constructing outside corridors on the west and north sides to keep the one thousand students comfortable during warmer months.[8] Thompson Electric Company, John Monier, and Mosel & Terrell Plumbing Company are subcontractors on the school plant, completed in 1950.

❖ ❖ ❖

Close by, the following year, 55,000-square-foot Sunset Ridge

Stores opens on a long stretch of North New Braunfels Avenue. The first of twenty-one tenants include a bakery, five-and-dime, beauty shop, cocktail lounge, and furniture store, impressing upon multiple media outlets to tout it as "San Antonio's largest and most modern community center." Hill & Combs is the general contractor for the $750,000 project. Mosel & Terrell is again a subcontractor along with Martin Wright Electric Company.[9]

❖ ❖ ❖

In 1951, Kelly Air Force Base on the city's southwest periphery awards Hill & Combs contracts totaling $923,873 to build twenty dormitories, along with two mess hall and administrative structures, which will accommodate an additional 500 to 750 airmen.[10] Future construction at Kelly, as it expands its logistics command after the war, will bolster the facility to become a major employer, especially of Mexican Americans.

❖ ❖ ❖

About the same time, G. W. Mitchell Construction takes on the challenging build of Trinity University's first dormitory on its new campus. The structure is erected at the base of a steep cliff located on the site of a former garbage dump and limestone quarry north of downtown.[11] The "hilltop" campus—the third and permanent location for the university—opens in 1952.

❖ ❖ ❖

Aptly narrated in a 2011 City of San Antonio video about the city's aviation history, "The 50s were definitely the time the airport began to soar." Construction takes off on a new terminal and administration building at San Antonio Municipal Airport with G. W. Mitchell Construction. In January 1952, the *San Antonio Express* proclaims the

$1.2 million project, which includes a wing dedicated to international travel, ahead of schedule for an October completion.[12] In the decades that follow, billions of public dollars will fund the expansion of the once-local air station into San Antonio International Airport.

The elaborate Alameda Theater opened in 1949 with considerable fanfare.

Architect Bartlett Cocke Sr. (far right and pointing)
at the future site of Trinity University, 1946

*Beginning of construction at Trinity University,
employing lift-slab construction, 1950*

Chapter Four
1953–1962

	Pearl Brewing Company Expansion Program
	University Presbyterian Church
Familiar	USAA Building
& Famous	St. Philips College Upgrades
Places	La Villita Assembly Hall
	North Star Mall
	McCreless Shopping City

In 1953, Dwight D. Eisenhower becomes president and during his two terms in office is credited with creating the US Interstate Highway System. The construction of highways in San Antonio helps to fuel its growth. Due to strong public support and post-war federal funding, expressway planning accelerates.[1]

The *San Antonio Express* reports "Many of the major construction projects now in progress in the San Antonio area are being performed by AGC members" and mentions the nineteen members of the San Antonio chapter.[2]

One member, F. A. Nunnelly General Contractor, builds on its good reputation with Pearl Brewing Company by completing a $468,464 expansion program that includes new cellars, a warehouse basement, and canning facilities.

That same year the newspaper announces the end of an eight-week strike when AGC and members of the local International Union of Operating Engineers reach a wage agreement, a prorated increase of fifteen cents per hour. Work resumes on sizable additions to Santa Rosa Hospital, the new Sears, Roebuck and Company department store on Southwest Military Drive, Southwestern Bell Telephone Company's main building, and various new schools within San Antonio Independent School District.[3] Walsh & Burney Company builds the addition to the hospital with Thornton Miller as superintendent, and W. S. Bellows Company is the contractor for the Southwestern Bell Telephone building. H. J. von Rosenberg constructs the $2 million Sears, Roebuck and Company store.[4]

❖ ❖ ❖

University Presbyterian Church's first dedicated house of worship reaches completion on Christmas Eve 1954, just four months after foundations are laid. It sits fittingly at the corner of Shook and Bushnell avenues adjacent to Trinity University, for whom the church was formed and where its congregation had gathered since 1950. Church records describe the sanctuary as "natural and functional in contemporary design and in the elevation and churchliness found in the best of the traditional." The builder, Walter Bowden, executes a design that best utilizes funds and materials available to bring the small church to life in time for its first worship service December 26.[5] Two years from now, Gene Treiber finishes construction of the education building.

❖ ❖ ❖

Well before *Texas Monthly* describes the 1975 United States Automobile Association (USAA) campus as one of the office wonders of the world and the second largest building of its kind, there is the smaller, yet stunning-for-its-time, Broadway location.[6] Henry C. Beck Company breaks ground on the second USAA headquarters building in 1954, with Mel Cox as superintendent, and completes the project in 1956. John F. Beasley Construction Company partners on the steel erection. The local chapter of the American Institute of Architects will describe it years later as a "spectacular 1950s period piece."[7] The numerous subcontractors involved include A. J. Monier & Company, Samuels Glass Company, Winn-Lee Masonry Company, and Paul Wright Electric Company.

❖   ❖   ❖

On August 3, 1954, *The San Antonio Light* reports, "The highly infectious building fever, which broke out locally last spring, reached the proportions of an epidemic early in the summer, affecting San Antonio and St. Philip's colleges most seriously."[8] F. A. Nunnelly General Contractor, exemplifying a major part of the outbreak, outbids a dozen leading construction firms to provide $699,300 in upgrades to the St. Philip's campus.[9] The college's oldest building, the Johnson Hall administration building from 1927, is demolished to make way for the construction of modernized administration and classroom buildings, the student union building, auditorium, and terraced walkways, marking "the apex of a long history that began in 1808 as a sewing class for girls."[10]

❖   ❖   ❖

In 1958, G. W. Mitchell Construction begins work on one of San Antonio's most innovative and uniquely engineered buildings to date. Designed by W. E. Simpson Company, La Villita Assembly Hall impresses with an inverted domed roof resembling a 132-foot bicycle wheel, supported by steel strands strung between rings. This is among the first cable-supported roof designs in the country.[11] The circular structure fulfills a dire need in the city for a convention and events facility and, upon its completion the following year, goes on to host countless proms, weddings, and public meetings.[12]

❖　❖　❖

Construction taking place in the late 1950s also satisfies an appetite for redefining community life and the shopping experience, not only in San Antonio but in other sprawling suburbs across the country. Federal dollars are now funding 90 percent of expressway construction costs, and San Antonio leads other Texas metros in carrying through with transportation infrastructure to support its businesses and shopping areas.[13]

South Texas is not far behind trendsetter Minnesota in embracing the enclosed shopping mall concept, a consumer phenomenon a world unto itself and free from harsh weather.[14] North Star Mall, built by D. J. Rheiner Construction Company, opens in September 1960 at San Pedro Avenue and Loop 13, a primary artery for the city and one of an abundance of expressways now aiding traffic flow. The mall provides sixty acres of air-conditioned, landscaped, one-stop comfort, with parking for 3,000 vehicles, and a dedicated bus stop and taxi stand.[15] Wolco Corporation was awarded the contract for prestressed concrete to build the 350,000-square-foot mega complex, and 800 local men were hired to move 360,000 cubic feet of earth on the $12 million project site.[16] Subcontractors include Prassel Manufacturing Company, Martin Wright Electric Company, A. H. Beck Foundation Company, Mosel & Terrell Plumbing Company, and A. J. Monier & Company.[17]

The *San Antonio Express and News* describes North Star's location as "the fastest growing section of America's seventeenth largest city," citing this northern part of town as where the explosive growth is occurring. Building San Antonio's first massive indoor shopping venue achieves more than just accommodating the migration of population away from the inner city. Shopping mall theory suggests another purpose: that neighborhood "climate-controlled monuments" serve as equalizers where "one moves for a while in aqueous suspension."[18] North Star's fifty-one stores, eateries, and businesses, enhanced by tropical gardens with resting benches and music piped in from KENS Radio Satellite Studio, provide carefree entertainment and otherworldly escape to the area's more than 44,000 families who might while away the hours shopping and dining, and even tend to personal investments at the Dempsey-Tegeler service booth.[19] A mixed bag of other tenants includes Zales Jewelers, Fox Photo, Luby's Cafeteria, and various apparel, furniture, toy, and gift shops.[20] During the grand opening, the Texas Store offers giveaways including a deer rifle, case of Champagne, and round-trip airfare for two to Acapulco.[21]

❖   ❖   ❖

New major mall construction is also happening in the southeast Highlands neighborhood. General contractor Guido Brothers Construction Company facilitates the development of McCreless Shopping Village into a much grander McCreless Shopping City with full-line department stores in 1962. Forty new stores are added to the existing twenty-two for a total of sixty-two retail options at what locals call "Big M City."[22] Gold coins are given out on opening day as helicopters drop play money redeemable for copter rides around the city.[23]

A few big-name tenants are celebrating company anniversaries this year including Sommer's Drug Store, in business fifty years and giving away Westinghouse electric roasters and preparing charcoal-broiled meals for its patrons.[24] The eminent Handy Andy Supermarket chain celebrates thirty-five years with the debut of No. 25, its most comprehensive store to date, offering a year's supply of bacon and two sides of beef as special prizes, as well as free coffee measuring cups and a pair of Vanderbilt 15 gauge nylon hosiery to the first 200 women who make a five-dollar purchase.[25] Those who trade with sixty-year-old Penney's get a special buy on a portable automatic phonograph with mahogany finish for forty-four dollars.[26] Wolco Corporation, Samuels Glass Company, and Nathan Alterman Electric Company contribute to the project's completion.[27]

The McCreless Mall is more than just a shopping center; it's a community gathering place where friends meet and plans are made, until its demolition in 2005.

Floyd Arthur Nunnelly Jr. built on the company's good reputation with Pearl Brewing Company.

USAA's Broadway location was built by Henry C. Beck Company in 1958.

Left to right: A. H. Cadwallader Jr., board chairman of Wolff & Marx; E. C. Sullivan, president of Wolff & Marx; D. J. Rheiner, D. J. Rheiner Construction Company; and architect Bartlett Cocke, Bartlett Cocke and Associates, at the construction site of North Star Mall

Chapter Five
1963–1972

Familiar & Famous Places	Inter-Continental Motors Frost Motor Bank and Garage Civic Center (San Antonio Convention Center) Tower of the Americas Texas Pavilion (UTSA Institute of Texan Cultures) HemisFair, Italian Pavilion HemisFair, Pearl Palm Pavilion HemisFair, Women's Pavilion Hilton Palacio del Rio South Texas Medical School Laurie Auditorium at Trinity University US Army Medical Field Service School

The year is 1963 and change is blowing in the wind. The Beatles release their debut album *Please Please Me*. Martin Luther King Jr. delivers his "I Have a Dream" speech on the steps of the Lincoln Memorial. In Dallas, President John F. Kennedy is fatally shot, and Vice President Lyndon B. Johnson is sworn in as the 36th president of the United States.

Months before Kennedy's assassination, the folks at Inter-Continental Motors Corporation in San Antonio are not thinking about The Beatles. They are focused on *Beetles*. Volkswagen Beetles, along with a full line of sedans, Kombi Station Wagons and Karmann Ghias to be showcased at the newly opened Volkswagen Sales and Service Center at 3303 Broadway. One of the country's most innovative and esthetically pleasing auto dealerships, built by Davis and Chandler Construction Company, occupies a verdant, three-acre site sheltered by mature oak and elm trees near Brackenridge Park. Design and construction were carefully planned to not interfere with the natural landscape. Utilities are installed underground, and the used car inventory is parked to the north of the building on padded "islands." At night, because of an intricate lighting system, the building seems to float in the air as special effects accentuate the colors of the different models. In addition to two polished showrooms, a lounge, and mezzanine overlooking a twenty-six-stall shop area, a drive-in service canopy diagnoses repairs on the spot with a special lift hoist not seen before in San Antonio.[1]

❖ ❖ ❖

Volkswagen owners are likely pleased about the 1965 opening of Frost Motor Bank and Garage and the convenience of drive-through banking, a major addition to downtown bringing six levels of parking and 49,000 square feet of basement facilities that include a lounge area for "restful" banking. The most unique feature of the project—a tunnel underneath Commerce Street connecting the new building to the main bank across the street—extends sixty-five feet long and eight feet below street level. The *San Antonio Express and News* depicts the construction as "a story of skilled engineering and experienced workmanship." Browning Construction Company and job supervisor James Parrott constructed the tunnel in atypical fashion to maneuver buried utility lines, completing it in sections so as not to interrupt daily traffic flow. Mechanical contractor A. J. Monier & Company combined the heating and cooling sources of the two buildings through the tunnel. L. E. Travis & Sons provided painting and wall covering in the Motor Bank, and Wm. H. LaDew installed sprinkler fire protection. Fred Clark Electrical Contractor and Samuels Glass Company also contributed to the project's completion and success.[2]

❖   ❖   ❖

The ground floor of the new motor bank displays a scale model of HemisFair '68 illustrating future buildings and attractions.[3] The six-month, international exposition will focus the world's attention on San Antonio and forever establish the city as a major tourist destination.[4] In January 1966, Governor John Connally announces the first and largest of some fifty HemisFair exhibitors, the State of Texas. Private and public money flow into the construction of plazas, pavilions, and an elevated monorail. That March, joint venture contractors Darragh & Lyda of San Antonio and H. A. Lott of Houston (Lyda-Lott) kick off fair construction with the Civic Center.[5] By October, the "dramatic social and civic hub" comprised of an arena, domed meeting hall for showcasing many of the fair's participants, and 2,800-seat theater that together absorb fifteen acres of downtown, is 30 percent complete and seven weeks ahead of schedule. After HemisFair, the center converts to city use for conventions and cultural events.[6]

In August, Lyda and Lott partner on the fair's centerpiece, the Tower of the Americas. Elmer Joiner acts as superintendent.[7] The challenging project entails ground-level construction of the tower's crown restaurant and observation decks, exceeding one million pounds, which are then meticulously lifted 622 feet and placed using steel rods.[8] The tower will remain an iconic symbol of San Antonio even as future developments surround it.

The same year, civic leaders reach an agreement to preserve twenty-two structures in Germantown—among the city's oldest neighborhoods settled primarily by immigrants—in the fair's final design.[9] This is far fewer than the San Antonio Conservation Society desires. Forty-three years later, volunteer members of the San Antonio Chapter of AGC will restore one such structure, the Koehler House, into offices of the San Antonio Parks Foundation.

Fair construction continues as Warrior Construction breaks ground on the $10 million state pavilion, the Institute of Texan Cultures, on February 20, 1967, with Curtis Lynge as project superintendent. In a statement, Governor Connally describes the state building as "one of the most significant projects ever undertaken in Texas," and "a first effort in our proud history to portray fully and accurately the fascinating cultural heritage which is ours."[10] In the weeks ahead, Lynge and Joiner will, with their respective projects, battle the local Laborers' International Union of North America through two strikes, as well as aggressive construction schedules to make up for lost work.[11] Post-fair, the Institute of Texan Cultures becomes a permanent collection of Texas history.

In *Bartlett Cocke General Contractors: The First 50 Years, 1959–2009*, Bartlett Cocke Jr. will recall how HemisFair "was an important learning and confidence-building experience." From 1967 to 1968 he works seven days a week erecting pavilions and a dedicated office for the frequently visiting governor.[12]

During the same stretch, F. A. Nunnelly General Contractor builds the Italian Pavilion and Pearl Palm Garden, the fair's first industrial exhibit. At the groundbreaking, hundreds of balloons containing free admission cards signed by Pearl board chairman and president Otto A. Koehler are released into the air.[13] Guido Construction builds the Women's Pavilion.

These are exciting times for the city and the industry, with $50 million in approved construction backed by enthusiastic participation from a growing list of nine foreign governments and eight prominent industrial exhibitors. The *San Antonio Light* projects, "There will be a building boom that will echo throughout the area for many years."[14]

However, not without setbacks. In his writings, Cocke Jr. will hark back to the spring 1968 opening when H. B. Zachry, one of the fair's organizers, requests a meeting with the construction firm owners. Zachry bears unwelcome news: "The fair was broke and we would receive no further payments. That was a gut-wrenching experience," Cocke recalls. These are equally unsettling times for the country with the deaths of Martin Luther King Jr. and Robert Kennedy, along with Vietnam War protests. Fair organizers later attribute lower-than-hoped attendance to this state of social unrest.[15] In lieu of monetary compensation, the more fortunate contractors receive cartons of HemisFair admission tickets.

Shifting forward fifty-four years to 2022, the *San Antonio Express-News* will report on a new mixed-use development at the former HemisFair location. The plan includes a park, a hotel, and apartments where fairgoers once strolled, sipping Pepsi Cola and Pearl beer. The article states, "The Zachry family has deep ties to Hemisfair and Hemisfair Park Area Redevelopment Corp., which is responsible for revitalizing the site. H. B. 'Pat' Zachry helped bring the world's fair to San Antonio and built the Hilton Palacio del Rio quickly to host its guests."[16] The reference is fitting. Back in 1968, H. B. Zachry Company completes the 500-room Hilton Palacio del Rio on March 30, in 202 days.[17] Workers assemble and even furnish the guestrooms off-site, and then hoist them into place at Alamo Street. This is a hefty achievement and the first of its kind in modular building.

❖ ❖ ❖

Continuing in 1968, San Antonio's first medical campus, conceived initially as South Texas Medical School, opens as the University of Texas Medical School at San Antonio on the site of the former Joe J. Nix Dairy Farm.[18] The $12 million project includes a four-story auditorium, library building, and both basic and clinical science wings.[19] Built by G. W Mitchell & Sons, it lays the foundation for the future South Texas Medical Center.

❖   ❖   ❖

G. W. Mitchell & Sons is also selected to build the 2,986-seat James W. and Dorothy Laurie Auditorium at Trinity University, the campus's 43rd major building and first large enough to accommodate the entire faculty and student body in a single sitting. Bringing central purpose to the structure achieves immediate past president James W. Laurie's vision of it as a "powerful instrument for unity." During the dedication ceremony on October 30, 1971, Dean Bruce Thomas says, "This building is not just the 43rd building on this campus, nor just the structure that completes the fine arts center, and its significance cannot be understood by simply referring to it as the culmination of the great building era."[20] Freeman Oates is the project superintendent. Forty-two years earlier, he worked as a laborer on the Atkinson Residence project (McNay Art Museum). In addition to housing the university's music, art, drama, and communications offerings, the auditorium becomes a respected cultural venue as the site for San Antonio Symphony Orchestra performances and is praised for its acoustics.[21]

❖   ❖   ❖

In late 1972, Browning Construction Company wraps up construction of the United States Army Medical Field Service School, comprised of two buildings for academics and administration. Subcontractors include R. L. Brand & Son and Redondo Manufacturing Company. Browning has two other major projects underway: the Aircraft Engine Inspection and Repair Hangar at Kelly Air Force Base and a large building for Southwestern Bell Telephone Company."[22]

Lyda-Lott built the Convention Center Arena, central to HemisFair '68.

*HemisFair '68 construction site with the Tower Life
and Alamo National Bank buildings in the background*

*The Tower of the Americas, pictured under construction,
remains an iconic symbol of San Antonio.*

*Gerald Lyda Sr. was a no-nonsense contractor
who helped shape the San Antonio skyline.*

Chapter Six
1973–1982

Familiar & Famous Places	Audie L. Murphy Memorial Veterans Administration Hospital Frost Bank Tower The University of Texas at San Antonio USAA Alamo National Bank Remodel San Fernando Cathedral Restoration HemisFair Arena Roof Raising San Antonio Marriott Riverwalk Hyatt Regency San Antonio One Riverwalk Place San Antonio Museum of Art

In 1973, while the Watergate scandal chips away at Richard Nixon's presidency, things in the Alamo City are more upbeat. The San Antonio Spurs make their basketball debut. Rod Stewart, Led Zeppelin, and Sonny & Cher perform in the relatively new Convention Center Arena. And long before H-E-B dominates the grocery market, Brand Names Foundation names locally based Handy Andy National Food Retailer of the Year.[1]

The construction beat goes on in San Antonio despite looming economic concerns. In the burgeoning South Texas Medical Center, general contractor J. W. Bateson Company completes the $47 million Audie L. Murphy Memorial Veterans Administration Hospital as the United States withdraws armed forces from Vietnam. The project represents the forefront of medical facilities, with comfort-focused rooms and recreation dens, sophisticated classrooms and laboratories, and a theater for stage performances. Exemplary design and construction, along with research and education offerings unprecedented among all 170 veterans administration hospitals, attracts national recognition. President Nixon sends the congratulatory message, "No community in America is more blessed with good medical facilities than San Antonio, and none has shown more energy and determination in acquiring them."[2] Handy Andy takes out a prominent advertisement in the *San Antonio Light* saluting veterans and the opening of the 760-bed facility.[3]

❖ ❖ ❖

That same year, Tom Frost Jr. officiates the dedication of another facility, a new home for Frost Bank and the largest commercial office property downtown. Rather than the traditional ribbon cutting, Mrs. T. C. Frost Sr. symbolizes the building's opening by unlocking it with a gold key. "Since we are guardians of several millions of your dollars, it would be more appropriate," Frost says.[4] The $20 million, twenty-story banking and office tower at 100 W. Houston Street sits adjacent to the existing Frost Motor Bank and Garage. The private Plaza Club will occupy its top floor.[5] Henry C. Beck Company is the general contractor. Wolco Corporation furnishes precast concrete. Samuel Dean Company also contributes to the project.

❖ ❖ ❖

Fourteen miles northwest of the downtown tower on the new

University of Texas at San Antonio (UTSA) campus, construction flour-
ishes on the institution's first seven buildings. But not without problems.
In February 1974, according to UTSA president Peter Flawn, the campus development is only 48 percent complete.[6] A May 1972 contract
between the University of Texas System and general contractor T. C.
Bateson Company (not to be confused with J. W. Bateson Company)
stipulates a 710-day construction schedule. Delays threaten enrollment
plans. To what and to whom these delays are attributable, aside from
inclement weather, is up for dispute. T. C. Bateson seeks restitution
claiming breaches by the owner, including furnishing the site in "faulty
condition," time-consuming design changes, and misrepresentation of
the amount of rock excavation—double of that originally agreed.[7]

❖   ❖   ❖

Inflation is on the rise, surpassing 11 percent in 1974. Dr. W.
Philip Gramm, professor of economics at Texas A&M University and
economic advisor to the United States, speaks to the Junior League of
San Antonio about government spending and runaway inflation—topics
that draw the attendance of H. B. Zachry, who will reiterate their impact
on the construction industry. Mrs. Bartlett Cocke Jr. is acting League
president.[8] Dr. Gramm will go on to be elected as a Republican to repre-
sent Texas in the US Senate in 1984.

After a year of construction delays, the Humanities-Business
Building at UTSA finally opens in 1974. The Science-Education and
Arts Building and Convocation Center follow.

The inflation rate falls to 9 percent the following year as public and private investment in construction projects fuels city growth.
Municipal construction projects alone are nearing $165 million and
are predicted to play a significant and ongoing role in job creation.[9] A
San Antonio Light front-page headline, "Economy Surges Upward,"
underscores how easing inflation is contributing to twenty-year record

gains. Below the headline appears a photo and mention of the opening of San Antonio's "new, ultra-modern" federal courthouse, known during HemisFair as the federal pavilion.[10] The contract to renovate the pavilion and an adjoining building was awarded to Lyda Inc. two years prior with Nathan Alterman Electric Company also involved.

❖ ❖ ❖

Mid-decade, a new corporate campus opens its doors in northwest San Antonio to headquarter the city's largest employer. Retired Brigadier General Robert F. McDermott presides over the 1976 dedication of USAA's massive complex that brings all 3,500-plus employees together in one place to work, dine, shop, bank, and exercise. Henry C. Beck Company is the general contractor of the five-building compound containing 3.1 million square feet of horizontal office space—second only to the Pentagon in size. Its construction took five years, requiring nearly 15,000 tons of steel and 87,000 cubic yards of poured concrete. Hall Sprinkler Company installs irrigation on the 286-acre site and natural area.[11] In 2017, the president of the San Antonio Economic Development Foundation, Mario Hernandez, will tell the *San Antonio Express-News*, "USAA helped put economic development on the map for San Antonio." It will be 2018 before H-E-B takes the lead as the city's dominant employer.

The construction of USAA, UTSA, and the South Texas Medical Center together spark housing development far from the city's core. But downtown is not forgotten. Renovations are underway and civic leaders look to new, suitable hotel rooms to accommodate the growing tourism and convention business.

Alamo National Bank, a distinctly recognizable feature of the downtown skyline since 1929, undergoes significant remodeling during this time. Superintendent Ross Coker with Bartlett Cocke Jr. Construction Company oversees a complete revamp from the basement to the eighth floor.[12]   To commemorate the occasion, the bank hosts receptions on five consecutive nights in 1976. The following year, Alamo National invites Cocke to serve on its board of directors. In this role, he develops valuable friendships that will benefit his growing company.

❖   ❖   ❖

By 1977, extensive restoration of San Fernando Cathedral by Guido Brothers Construction Company is progressing. The three-phase bicentennial project focuses on the eighteenth-century apse and nineteenth-century nave and includes adding a parish center, sacristy, and priests' residence, the latter on a former ALLRight Parking lot in Main Plaza.[13] Louis Guido Jr. serves as president of the San Antonio Chapter of AGC. By 2019, five company representatives will hold this leadership position.

❖   ❖   ❖

Meanwhile, City Council approves appropriating $4 million to raise the roof of HemisFair Arena to accommodate 6,000 more specta-tors. G. W. Mitchell & Sons, general contractor for the remodel, deploys more than thirty-eight hydraulic jacks to accomplish the thirty-three-foot goal.[14] It take cousins Doug and Gary Nunnelly, ironworkers with subcontractor Texstar Construction Corporation, three days, lifting the 2,260-ton roof inch by inch. "The jacks had to move simultaneously," Gary Nunnelly later recalls. "Not one could be off, so we ran around continuously checking the jack the whole time." Few comparable roof projects have preceded that of the arena, and completing a highly com-plicated task of this nature, on schedule, earns G. W. Mitchell the Texas Building Branch–AGC Outstanding Construction Award in 1980.

❖   ❖   ❖

San Antonio's segue into hospitality and tourism gains momentum during the late seventies and early eighties with the completion of two major downtown hotels. General contractor W. S. Bellows of Houston finishes work on the San Antonio Marriott Riverwalk in 1979. Challenged by the small site located on an extension of the River Walk, crews had no place to build but up. The skyscraper's thirty stories stretch 350 feet. Featuring 500 guestrooms and 10,000 square feet of ballroom space near the River Walk and San Antonio Convention Center, the modernistic skyscraper is exactly what civic boosters are looking for as San Antonio chases more tourists and conventioneers. The project boasts landscaped terraces, an indoor-outdoor pool, and a sloping glass atrium at the fifth level.[15] Nathan Alterman Electric Company is a subcontractor.

❖ ❖ ❖

With continued public investment, Hyatt Regency San Antonio opens in 1981 at a price tag of $38 million, adding an additional 631 guestrooms to downtown. Reporting for the *Sunday Express-News,* Mike Greenberg writes, "No San Antonio building project has attracted as much public interest as the new Hyatt Regency hotel and the Paseo del Alamo, the lush water garden that links the River Walk and Alamo Plaza through the hotel's lobby." Atlanta-based Hardin International is the general contractor and G. W. Mitchell & Sons erects the concrete frame. Construction of the sixteen-story, C-shaped high-rise entails jackhammering narrow grooves in the concrete exterior to create a richly decorative texture. Accessed from the hotel's atrium lobby, The Landing Jazz Club will become the home of the Jim Cullum Jazz Band and its successor the Happy Jazz Band up until 2012. Radio broadcasts from the club find a national jazz audience and further spread River Walk allure.[16]

❖ ❖ ❖

Bartlett Cocke Jr. Construction Company builds eighteen-floor One Riverwalk Place for commercial office use in 1981. Coker again is the superintendent and according to Cocke Jr., "did a superb job." In documenting the company's first fifty years, Cocke will remember laborer Big John "insisting that as each floor was poured, the concrete set up faster because we were getting closer to the sun. I didn't have the heart to tell him the sun is 93,000,000 miles away." [17] In 2013, USAA Real Estate Company will acquire the 261,633-square-foot office building. [18]

❖   ❖   ❖

The same year, project manager Tom Guido is keenly focused on the adaptive reuse of the old Lone Star Brewery, built in 1884. Its conversion to the San Antonio Museum of Art—one of Guido's first such projects in San Antonio—tells a story of skill, determination, and $7.2 million in public and private investment. The project, under the watch of superintendent Jim Perkins, poses difficulty from the outset. The structure is in a state of disrepair. Floor levels are uneven, cast-iron columns and steel beams rusty, and freight elevators and stairs unusable. What exists is a magnificent, ornate brick exterior with well-proportioned and varied interior spaces. Guido's wife and business partner, Maryanne Guido, will in later years cite the complex undertaking as helping to change the trajectory of the family business and establishing Guido Brothers Construction Company as specialists in historical renovations.

The museum opens to the public in March and the company is awarded the Outstanding Construction Award from the Texas Building Branch–AGC in 1982.

The University of Texas at San Antonio under construction in 1974

*The University of Texas at San Antonio president Dr. Peter Flawn
expressed concerns about campus construction delays.*

*Aerial view of The University of Texas at San Antonio
under construction, 1976*

*Converting the old Lone Star Brewery into the San Antonio
Museum of Art established Guido Brothers Construction
as specialists in historical renovation.*

Henry C. Beck constructed five buildings
on the USAA compound in 1980.

*The downtown Marriott hotels
further elevated San Antonio as a tourist destination.*

Chapter Seven
1983–1992

	Interfirst Bank Plaza (Bank of America Plaza)
	Pace Foods (Lucifer Lighting Company)
	H-E-B Corporate Headquarters
Familiar	Palo Alto College
& Famous	Hemisfair Park Redevelopment
Places	Rivercenter Mall
	San Antonio Marriott Rivercenter
	SeaWorld San Antonio
	Marketplace H-E-B, Bandera Road
	Fiesta Texas (Six Flags Fiesta Texas)

John Forgy of Forgy Construction Company is president of the San Antonio Chapter of AGC in 1983. He and fellow member Lane Mitchell of G. W. Mitchell & Sons serve on a building committee. Like his grandfather back in 1932, Mitchell is active and engaged with the association. The goal of the committee is to build a new chapter office near the airport. Far more than brick and mortar, the office building would serve as a hub—an important gathering place for the city's leading commercial contractors.

Almost forty years later Mitchell recalls, "John was a good guy, a union contractor as most of us were back then and was known to do quality work. One of his right-hand men was John Malitz, who I met through AGC.

"Malitz left Forgy shortly after the AGC building was built to start his own company. He never came back to AGC. Ironically, John told me he had planned to transfer Forgy Construction over to Malitz as he had no interested heirs."

As with some other legacy construction companies in the years ahead, there would be no heir apparent. Yet, the work of building the city carries on.

Downtown, Henry C. Beck Company wraps up construction of Interfirst Bank Plaza, whose twenty-eight stories broke ground in 1981 as First International Plaza. Through mergers and acquisitions, the building will wear many names. Architecture critic and historian John C. Ferguson writes in *Texas Architect,* "The strengths of Interfirst Plaza make it a building of major significance in both architectural and the commercial development of downtown San Antonio."[1] Rising 387 feet as one of the tallest in San Antonio's skyline, it contributes more than a half-million square feet of high quality, Class A office space to the city's business district.[2] Moreover, its distinctive pleated and stepped-back exterior easily make it one of the most recognized buildings in the city.[3]

❖ ❖ ❖

Also in 1983, Guido Brothers Construction completes the headquarters and manufacturing facilities for the renowned Pace picante sauce.[4] An anomaly in the realm of food processing plants, the southwestern style edifice is a people-focused space with shaded courtyards and colorful Mexican folk art within, nestled on seven scenic acres near Salado Creek. Subcontractors include Crown Steel, Samuel Dean Company, and L. B. Palmer & Sons. Project manager Bob Walker later remembers the unique project as "one of the early logistics-centric operations that I experienced, designed to facilitate the flow of trucks in and out with the produce deliveries to the back and empty jars in and full jars out to the side."

The $1.1 billion sale of Pace Foods to Campbell Soup Company years later will be both a source of sadness for San Antonio when salsa operations relocate out of the city, and one of epic good fortune as proceeds help fund the metamorphosis of the old Pearl Brewing Company site into a prosperous, nationally recognized mixed-use space.

❖ ❖ ❖

Food is a reoccurring theme in 1984, as distinctive work is being performed at the site of H-E-B Corporate Headquarters. Once the location of a military arsenal and then a "hopeless jungle of blight on the southern edge of downtown," the storied grounds are now being transformed.[5] Bartlett Cocke Inc. is the general contractor. Looking back, Bartlett Cocke Jr. recalls, "There was new construction, lots of remodeling, and incessant change orders." He doubted the project would be completed on time.[6]

Jamey Arnold is acting superintendent. He is twenty-six years old, and this is his first big job. The pressure is on to keep to schedule, and an abandoned structure next to an old horse stable stands in the way. Arnold decides it needs to go and directs Mark Cuppetilli of M&M Demolition to take the building down early on a Saturday morning. A stranger lingering outside the construction fence asks to come inside and watch. Arnold refuses but invites the gentleman to safely observe the demolition from outside the fence. The following Monday morning, Arnold is summoned to a meeting with the architects, consultants, and corporate executives, where he is chastised for demolishing the building. The stranger from outside the fence, present at the meeting, stands up in Arnold's defense, blaring, "He's just the kind of guy we need on this project!" That stranger is H-E-B Chairman Charles Butt.

Beth Standifird, a Master of Science in Architectural Studies candidate at The University of Texas at Austin, tours the headquarters site, years before she becomes the librarian for the San Antonio Conservation Society Foundation. H-E-B's choosing to embrace preservation practices while making the design of compatible new construction impresses her.

"What I learned was that adaptive use also involved compromise—the judicious loss of some historical building fabric to preserve and revitalize the greater whole," she says. "But the result confirmed to me that I wanted to play a role in making more projects like that happen."

Judd Plumbing, Mesa Equipment Company, and Nathan Alterman Electric Company contribute to the project, which does in fact finish on time in 1985.

❖ ❖ ❖

Another project with a demanding schedule opens as planned in 1987 after receiving voter support in 1984. Alamo Community College District approves 108 brush-covered, rattlesnake-infested acres located at Interstate 410 and Palo Alto Road for its newest college, aptly named Palo Alto College. Construction is awarded to Kunz Construction Company. Company owner Harold "Bubba" Kunz, a former bricklayer described by employees as "tough and a construction genius," is up for the task. Mike Cooney, an estimator for Kunz, is also intimately familiar with the challenge that constructing an entire college campus within 468 days will entail.

"Needless to say," Cooney recalls, "I was stunned, shocked, and downright scared when Mr. Kunz walked into my office early one morning and stated, 'The liquidated damages on the project are $10,000 per day, and I'm sending you out there full-time as project manager to make sure that we do not have to pay a dime.'"

Encompassing eleven buildings, including a 15,000-square-foot library, theater, and classrooms for arts, science, and robotics programs, Palo Alto College promises higher education opportunities for San Antonio's growing South Side.[7]

❖ ❖ ❖

Kunz Construction goes on to complete the $10 million Hemisfair Park Redevelopment project, a desperately needed rejuvenation that commemorates the world fair's twentieth anniversary in 1988. Mindful of the Tower of the Americas as the park's central element, Superintendent Ronnie Hargett and Project Manager Andy Koebel link walkways, landscaping, and continuously circulating water features that radiate from the Tower to historic La Villita. Redevelopment also includes the construction of the "River Link" which connects the river level to Market Street in front of the convention center. Work involves demolishing existing concrete retaining walls as well as constructing earth retaining soldier piers and stone veneered cast-in-place concrete stairs and planters.[8]

❖ ❖ ❖

The river remains a focus in 1988 when the $200 million Rivercenter Mall and hotel duo opens, elevating downtown San Antonio even more as a tourist destination. The River Walk is further extended to form a lagoon around the glass-fronted, three-level mall and its more than one hundred retail shops, five sit-down restaurants, and two movie theaters.[9]

Bringing in a new hotel, however, meant out with the old. To escape destruction, the seventy-nine-year-old boutique Fairmount Hotel had to be relocated three years earlier to make way for the thousand-room Marriott Rivercenter. The herculean task was entrusted to Alamo Architects and Emmert International Hauling. To the delight of onlookers, city leaders, and the Conservation Society, they moved the 3.2-million-pound hotel from the intersection of Bowie and Commerce streets to La Villita, saving it from demolition.[10] With the Fairmount safely deposited in its new location, and the aid of a $15.7 million federal Urban Development Action Grant, mall and hotel construction could begin.[11]

Photos from a digital account in the *San Antonio Express-News* reveal a sign of the times—a boom in hospitality and the beginning of labor shortages within the construction industry. United States Immigration and Naturalization Service agents search the construction site for undocumented workers.[12] Marek Brothers and Nathan Alterman Electric Company contribute to the construction of the hotel. Manhattan Construction Company builds the adjacent mall. After completion, a river taxi carries Mayor Henry Cisneros, members of City Council, and other dignitaries to opening ceremonies and a browse through national grande dame Lord & Taylor's first department store in San Antonio.[13]

❖ ❖ ❖

Construction and hospitality fervor continue in 1988 with San Antonio's first major theme park. Texas Trident is the third-party construction manager at SeaWorld San Antonio on the Far West Side. Don Daughtry, representative for the firm, is not coping well under pressure. Remembering Daughtry, Bartlett Cocke Jr. writes, "He would scream, stomp, and get so red-faced that we thought he would never survive to opening day."

Bartlett Cocke Jr. builds the Penguin Encounter Exhibit and animal care facilities as well as the Ski Lake Grandstand. Randy Pawelek acts as project manager, Jamey Arnold as superintendent, and Earl Noble is a union carpenter foreman onsite. Nathan Alterman Electric Company is a subcontractor. For its work on the penguin exhibit, Cocke receives the Outstanding Construction Award from the Texas Building Branch-AGC.

❖ ❖ ❖

Forbes names H-E-B the thirty-sixth largest private company in the United States this year, and in 1990 Joeris General Contractors completes Marketplace H-E-B. This is Joeris' first H-E-B project, a milestone for the general contractor and a new store concept for the rapidly expanding grocer. James Lynch acts as superintendent, and subcontractors include Eldridge Electric Company and A/C Technical Services.

"It was their first store on such a large scale," says CEO Gary Joeris looking back "This project opened up an entirely new vertical market for our company and allowed us to create an expanded growth trajectory based on project diversification."

H-E-B would, too, grow. By 1995 the company will have amassed 224 stores, all in Texas.[14]

❖   ❖   ❖

San Antonio's relentless focus on growing its tourism continues in 1992 with the opening of its second theme park, Fiesta Texas, built on the site of another depleted rock quarry. The development team includes USAA Real Estate Company, a subsidiary of USAA. Lyda Inc. and Manhattan Construction Company serve as the two general contractors in a joint venture in charge of the construction, with Nathan Alterman Electric Company a subcontractor. Post-retirement, Carl Weyel, with Alterman since 1967, will remember how "everybody was behind, but the opening date never changed," and how installing underground electrical utilities required crews "to remove boulders as big as a car." Plus, he wryly adds, Lyda "jacked us around on the change orders."

The following year, Manhattan is bestowed the national Associated Builders and Contractors Excellence in Construction award for its work at Fiesta Texas.[15]

Carl Weyel with cigar and microphone at an AGC Safety Fair and Barbecue Cook-off. Regarding the Fiesta Texas project, Weyel said, "Everybody was behind, but the opening date never changed."

Harold "Bubba" Kunz and Mike Cooney faced a demanding schedule with Palo Alto College.

*Bartlett Cocke Inc. built the
H-E-B Corporate Headquarters.*

*Lyda Inc. and Manhattan Construction Company were
general contractors in a joint venture building Fiesta Texas.*

Chapter Eight
1993–2002

Familiar & Famous Places	Alamodome
	Nelson W. Wolff Municipal Stadium
	Central Library
	City Council Chambers Renovation
	San Pedro Springs Park Renovation
	McCombs Residence Hall, University of the Incarnate Word
	SBC Center (AT&T Center)

The Cold War ends in the early 1990s as does the justification for unprecedented defense spending. Significantly, the 1993 federal Base Realignment and Closure (BRAC) Commission recommends shuttering Kelly Air Force Base, but prolific local support temporarily saves it. BRAC again decides in favor of closing Kelly in 1995. This time it sticks. The loss of the prominent air logistics center, a major employer and home to generations of workers, is a gut punch to the local economy.[1]

The year 1993 is also when city leaders celebrate the opening of a multi-use, sales-tax supported sports facility called the Alamodome. Situated on the southeastern fringe of downtown, it represents the largest urban development since HemisFair.[2]

The $183 million project overcomes controversy surrounding

contaminated dirt, and the completed dome resembles to some critics a dead, upside-down armadillo. However, Mayor Nelson Wolff declares the new building—with pedestrian plazas, suspension cables, and four 300-foot concrete towers—a triumph, delivered under budget and on time.

Although the stadium fails to attract a National Football League team as former Mayor Henry Cisneros and others had hoped, it does become the home of the San Antonio Spurs when Wolff, in 1994, convinces city council to demolish the team's present residence, a mostly vacant HemisFair Arena, to make room for a convention center expansion. It also serves as an all-around "excellent facility with great sight lines, 65,000 comfortable seats, large suites, wide corridors, and a good sound system."[3] A number of contractors involved in building the dome include Lyda Inc. and Kunz Construction Company, responsible for lowering Montana Street to facilitate a new bus terminal for getting people to and from events, Martin Wright Electric, Moore Erection Company, Glenn Graham Plumbing, and Olmos Equipment Company.[4]

❖ ❖ ❖

Nelson W. Wolff Municipal Stadium opens the following year in April, just in time for the first pitch of the first baseball game between the San Antonio Missions and El Paso Diablos. The $8.3 million project, eight miles west of downtown, had to be completed in six months. Bartlett Cocke General Contractors pulled it off working two crews, seven days a week, sometimes eleven-hour days.[5] Cocke acknowledges the lineup of Superintendent Darrell White and City of San Antonio Project Manager Jim Mery for meeting the mark. T&D Moravits, Martin Wright Electric, and Moore Erection Company were also critical to accomplishing the 194-day build. The 6,400-seat stadium's open-trussed metal roof suggests industrial garage, while sloping berms and grassy areas say public park. Two clad stucco towers at the main entrance are a nod to neighboring Mission Concepción. It receives an Outstanding Construction Award from the Texas Building Branch–AGC in 1995.

❖   ❖   ❖

On a sweltering day in May that same year, an unmistakably different-looking building opens downtown at the corner of Soledad and Navarro streets. General Contractor H. A. Lott completed the "enchilada red," 238,000-square-foot Central Library with the support of $28 million in voter-approved bond funds and an additional $10 million from private sources and the city's general budget.

Hardly shy, the city's showy, $38 million new main library brings bold, geometric Mexican design to the site of a vacated and later demolished Sears store, earning it the Mind Science Foundation's 1995 Imagineer Award.[6] *The New York Times* reports one architecture critic's praise of "an instant international landmark" and another disparaging it, although cleverly, as "the loudest library in the country." Residents are equally at odds about world-renowned architect Ricardo Legoretta's hallmark use of explosive color that speaks more of Mexico than San Antonio.[7]

Sixteen years after the opening, Wolff, Bexar County judge during this time, recalls, "Any time you do anything striking, you'll receive a lot of criticism."

Showiness aside, the structure is functional, refreshing, and somewhat famous. In addition to the requisite books (more than a half million volumes), state-of-the-art information technology, and public meeting spaces, one finds acequia gardens, loggia-style terraces, and inviting courtyards complementing a six-story atrium. The library will host field trips for young and old, appear in the film *Selena,* and serve as fictitious espionage headquarters in *Spy Kids.*[8]

❖   ❖   ❖

M. J. Boyle General Contractor is awarded the low-bid contract to transform the hallowed Luby's Cafeteria at the corner of West
Commerce and Main streets into new City Council Chambers. The comfort food institution had moved into the former Frost Bank lobby back
in 1975 after Tom Frost Jr.'s gold-key opening of the new bank tower
across the street. The Luby's location served up loads of LuAnn Platters,
fried fish squares, and melty macaroni and cheese until 1989 when the
city purchased the building and in 1992 voted to relocate there.[9]

In 1994, company principal Mike Boyle is enthusiastic about renovating the ground floor of the renamed Municipal Plaza Building into a
new space where public officials will deliberate plans for city growth, but
he is aware there are challenges. The project's scope entails removing a
supportive concrete column, relocating a historic staircase, and restoring
a structurally unsound ceiling. "Many of my competitors were afraid of
the scope," Boyle will later recall. They had sound foresight. "The design
firm, RVBK, split up during the project. Beaty, the 'B' left the firm and
took the management of this project to his new firm."

Overcoming these difficulties, M. J. Boyle wins an Outstand-
ing Construction Award from the Texas Building Branch–AGC in 1995.
Boyle credits the entire construction team for the recognition, especially
Big State Electric, J&R Tile for the intricate tile and terrazzo work, and
Jary and Associates for ceiling art.

❖ ❖ ❖

Later that year, the renovated City Council Chambers is the location of a press conference concerning the growth of the local chapter
of AGC. Chapter President Boyle, Chapter Vice President Duane Pozza
of Bartlett Cocke, and Chapter Executive Vice President Doug McMurry join Mayor Wolff to announce a merger with the South Texas Chapter
in Corpus Christi. AGC leaders believe the new, larger group reflects
broader market opportunities for a thriving industry.

❖   ❖   ❖

Sensing opportunities to make improvements, city voters pass a $54 million bond initiative for parks, which includes funding a master plan to rehabilitate historic San Pedro Springs Park, a vital part of San Antonio's rich history since the 1600s.[10] Plans in hand, Kunz Construction works from 1997 to 1999 on the focal point of the restoration, replacing the existing swimming hole lakebed with a circulating swimming pool that attracts recreation seekers while also preserving ecological integrity, in particular the native cypress trees. The flow of the springs across the footprint of the new pool presents a challenge. Crews discover a concrete slab four feet below the elevation of the new pool, making traditional foundation preparation impossible. To keep the water level sufficiently below the pool bottom while concrete is placed, they install several different pumps on the downhill side of the water flow that allow construction to proceed. At nearby McFarlin Tennis Center, the team installs a new performance court as well as a rock and concrete circular entry that refreshes the 1954 facility. Project Manager Andy Koebel and Superintendent Rodney Harrell guide the improvements along with subcontractors Design Electric, Lopez Concrete, and Mission Plumbing. The restored park opens in 2000 and receives an Outstanding Construction Award from the Texas Building Branch–AGC.

❖   ❖   ❖

Institutions of higher learning in San Antonio continue to grow and evolve, and commercial builders are literally providing the foundation for the city's pursuit of academic excellence. Work performed to date at The University of Texas at San Antonio, Trinity University, and the Alamo Colleges exemplifies such progress, which gains more momentum in 2000 when Joeris Inc. completes the first of five projects at the University of the Incarnate Word's Skyline campus. McCombs

Residence Hall, Parking and Banquet Center offer upperclassman apartment-style housing and 13,500 square feet of ballroom space in what is considered the upper campus, dramatic and highly visible to San Antonians as it fronts northbound Highway 281. Project Manager John Casstevens and Superintendent Dale Nieder are instrumental in the construction of this first Skyline job, which sets the stage for future expansion and enrichment of student life.

Reflecting on the first of several projects Joeris will complete for the university, CEO Gary Joeris echoes the company's mission—to transform people and places.

"The people of the university have entrusted Joeris to transform the campus much as the people of San Antonio and around the globe trust the university to transform the future by educating students."

❖ ❖ ❖

As with HemisFair Arena, the Alamodome, and Nelson Wolff Stadium, SBC Center is a sizeable building project that begins with a plan and a promise. It is a promise to a city, to a sports team, and to community activists.

After the Spurs win the NBA Finals in 1999, civic and county leaders hatch a plan to use hotel, motel, and car rental taxes, instead of sales or property taxes, to build a new venue for the team and the San Antonio Stock Show & Rodeo. The Spurs agree to help pay for the construction of the East Side arena, and the Saddles and Spurs campaign launches. Establishing standards for bolstering participation from local minority- and women-owned construction companies is an underlying component, necessitated in the aftermath of complaints expressed during the building of the Alamodome that not enough had been done to include such firms.

Campaign coordinator Anne Whittington describes how "key campaign individuals spent many hours personally contacting and reaching out to Minority and Women-Owned Business Enterprises (MWBEs), making solid commitments to include them in the construction project."[11] Such businesses are promised 20 percent of the work.

On election night, a determined Bexar County Judge Cyndi Krier stresses that now the county must "build the arena and fulfill our promises."[12]

After initially signaling support for the joint venture team of Turner Construction Company and Browning Construction Company, the Spurs reconsider and instead select Hunt/SpawGlass as the general contracting team for the $175 million, 18,500-seat arena. They hire Joe Linson, former president of the Alamo City Chamber of Commerce, to coordinate MWBE participation.[13] Bexar County in turn hires a full-time coordinator, Renee Watson, and reconvenes an advisory committee to scrutinize every construction dollar spent.[14] Excluding sole-source suppliers and contractors, the effort achieves more than 30 percent MWBE participation, surpassing that of any large construction project owned by a Texas county and serving as an example for future projects. Companies working on the project with minority- or women-owned certification early on include Foster CM Group, Poznecki-Camarillo & Associates, Arias & Kezar, and Cortes Contracting. Linson tells the *San Antonio Business Journal*, "This has set a precedent," signaling that "in San Antonio, we have proven on a large, high-profile project that these firms can compete."

Olmos Construction, mechanical contractor Todd-Ford, CFS Forming Structures Company, and Nathan Alterman Electric Company help build the arena, with JAG Contractors, Alpha Insulation, and Richard's Rebar Placing playing primary roles. SBC Center opens in 2002.[15]

❖   ❖   ❖

Meanwhile, Christopher "Kit" Goldsbury, former president of Pace Foods and subsequent founder of equity investment firm Silver

Ventures, sets an example for diversifying future development and construction in San Antonio. The firm purchases the twenty-two-acre Pearl Brewery complex and develops a master plan for both revitalizing the area and preserving its historic structures.

*The $183 million Alamodome represented the largest urban development
since HemisFair.*

H. A. Lott built the unmistakably different Central Library.

*Bill Mitchell, Duane Pozza, Mike Boyle, and Mayor Nelson Wolff
in the renovated San Antonio City Council Chambers*

*Joeris Inc. built the first five projects of
University of the Incarnate Word's Skyline Campus.*

*Gary, Leo, and Karl Joeris played an important role
in the growth of San Antonio.*

Jan and Mike Boyle with Bexar County Judge Cyndi Krier.
During the campaign to build the SBC Center, Krier stressed
that Bexar County "must build the arena and fulfill our promises."

Chapter Nine
2003–2012

Familiar & Famous Places	Toyota Manufacturing, Texas
	Children's Cancer Research Institute (Greehey Children's Cancer Research Institute)
	City of San Antonio Animal Care Services Main Campus
	Chapel of the Incarnate Word
	Main Plaza
	US National Security Agency Texas Cryptology Center
	Senator Frank L. Madla Building, Texas A&M University–San Antonio

After decades of economic development focused on tourism, hospitality, and sports stadiums, big-time manufacturing comes to San Antonio in early 2003. Having courted Toyota Motor Corporation for nearly one of those decades, city leaders are euphoric when the company selects a site south of downtown for its $800 million, 1.8 million-square-foot Tundra pickup assembly plant. Mario Hernandez, president of the San Antonio Economic Development Foundation, tells news sources, "We had never made it to the final list for an auto plant."[1] Judge Nelson Wolff adds, "In fact, most of our efforts to attract any kind of manufacturing had resulted in failure."[2]

The successful Toyota deal contains $133 million in incentives, including land, tax abatements, a rail district grant as well as job training and utility infrastructure.[3] And not lost on the Japanese multinational company in its selection criteria—Texas is the largest buyer of pickups in the country.

For San Antonio, the deal means engaging more than 2,000 local trade workers and employing 2,000-plus people when the plant is up and running as potentially the largest of any project in Bexar County's history.[4] San Antonio AGC takes the lead in promoting local firm participation in building the plant, and in June, Bartlett Cocke General Contractors announces a joint venture with Detroit-based Walbridge Aldinger, the "absolute industry leader in automotive manufacturing plant construction," to perform most of the foundation work.

"We've worked hard for nearly a year to put ourselves in this position," says Bartlett Cocke executive Duane Pozza.[5]

It turns out to be a good decision for those who understand the San Antonio marketplace and for those who don't. While Walbridge Aldinger knows little about the city, the firm does hold an essential advantage—it knows its way around unions.

Controversy had been brewing for months about the possibility of Toyota imposing a union-only project labor agreement and how that would affect a majority nonunion construction community. A *San Antonio Express-News* front-page headline reads, "S.A. contractors worry about the possibility of a union-only deal."[6]

Advocating for the local industry by noting that 95 percent of companies choose not to engage with unions, the San Antonio Chapter of AGC strongly discourages Toyota officials from pursing an all-union pact. Judge Wolff weighs in, "Our concern is that people who live here will have an opportunity to work on the project."[7]

The automaker ultimately acknowledges San Antonio's nonunion realities and settles for a wage agreement allowing nonunion contractors equal footing in competing for contracts. This marks a sharp departure from Toyota's previous union-governed policies for US projects.

In addition to the Bartlett Cocke and Walbridge Aldinger partnership, San Antonio concrete reinforcing company Richard's Rebar Placing and Ohio-based Baker Concrete join forces to complete the remaining foundation work.[8]

By 2019, Toyota will employ 3,200 workers at the facility and will announce a $391 million expansion.[9]

❖ ❖ ❖

Even with the success of Toyota, Bartlett Cocke Jr. considers the Children's Cancer Research Institute, completed in 2003, the most significant facility his company has ever built. Significant not due to size or profitability, but because the scientists who work there will search for a cure for pediatric cancer.[10]

Jerry Hoog is the project manager. The job happens to be the last and a memorable one for Superintendent Earl Noble, on the brink of retirement after 27 years of helping to build places all over the city.

The $42 million institute, part of The University of Texas Health Science Center at San Antonio, is one of only two in the country dedicated exclusively to pediatric cancer research. It will later take on the name Greehey in receipt of a $25 million donation from The Greehey Family Foundation and become a forerunner in cancer genomics, DNA repair, RNA biology, and drug development.[11]

The 100,000-square-foot project receives the 2003 Associated Builders and Contractors Excellence in Construction Award for innovation and high-quality workmanship.

❖ ❖ ❖

In 2007, general contractor F. A. Nunnelly completes 151 Main Campus Animal Care Center, the hub of operations for the City of San

Antonio Animal Care Services. The original contract amount was $11.5 million, but the architect, subcontractors, and general contractor teamed up to save about $1 million. Project Manager Gary Christensen and Superintendent Scott Sutherland are part of the collaboration.

The West Side project consists of eleven pre-engineered metal buildings on fourteen acres, each with dedicated functions. Services include adoptions, owner reclaim, permitting, microchipping, stray drop-off, and mandated rabies observation.[12]

The same year, the San Antonio Area Foundation spearheads a no-kill initiative, distributing donated funds to nonprofits that offer spay/neuter surgeries and education programs. In part, because of the foundation and the excellent work at the center, San Antonio will achieve the national no-kill standard in 2015: more than 90 percent of healthy, treatable animals released alive and well.[13]

❖   ❖   ❖

In 2007, M. J. Boyle General Contractor adds another significant renovation to its project portfolio, that of restoring the Chapel of the Incarnate Word, one hundred years after its original construction for use by the Sisters of Charity who arrived on the scene in 1869.

Substantial updates to the chapel include renovating the basement into office and archive space and constructing a museum on the first level. Boyle's team also replaces the old organ with a new Schoenstein pipe organ and installs an 8,000-pound marble altar flown in from Italy. The original wooden pews are replaced with new, replicated seating. Even the iconic exterior steeple is repaired and returned to its original glory, a project in itself. For years, the beautiful bell tower, with its brick and four trumpeting angels, was the highest structure in San Antonio.[14]

J & R Tile, L. E. Travis & Sons Painting, and Corbo Electric Company are part of the team. M&M Marble and Guarantee Plumbing & A/C also play a key role in the $4 million project.

❖ ❖ ❖

As with the initial stages of the Toyota plant, civic debate is heating up over Mayor Phil Hardberger's plan to redevelop Main Plaza in 2006. The dust-up this time is not about labor unions, but rather the closing of streets in the heart of downtown.

A strong advocate of parks and public places, Hardberger embraces an ambitious vision for the cramped plaza. He wants to double its size, build a new fountain, and create a grassy slope between San Fernando Cathedral and the San Antonio River channel.[15] The mayor tells the *San Antonio Current*, "It will set us apart from the rest of the United States with a true European-style plaza that would be a great draw for tourists and locals."[16]

There is formidable opposition to the estimated $10 million project. Banker and downtown business leader Tom Frost Jr. says he is "unalterably opposed." As senior chairman of Cullen/Frost Bankers, which owns nearby Frost Bank, Frost worries blocking off streets will devalue the property he opened with gold-key pride thirty-three years earlier.[17]

Another naysayer is Bexar County Commissioner Tommy Adkisson, whose precinct includes downtown. His opposition is important because, if given the green light, part of the redevelopment funding may have to come from county coffers.

After months of controversy, Hardberger and his allies, including AGC, cathedral rector Father David Garcia, and the Main Plaza Advisory Board, prevail, and the fast-tracked project is approved.

Kunz Construction Company is awarded the job and very quickly encounters uncharted obstacles at the 275-year-old location. Underground, they discover burial sites and historically significant items requiring attention and fear significant delays, or worse, a halt to the project. After sorting through the findings with city officials, crews are able to continue installing massive slabs of stone paving and pathways that meander around fountains and rectangular pools. Similar to the 1988 work in Hemisfair Park, they also connect the cathedral to the River Walk link at Portal San Fernando.

The $12.6 million project is completed in 2008 and receives accolades from a happy Hardberger, who offers his personal thanks for the hard work and dedication put into the redevelopment. "It has been a tremendous team effort and much of it would not have happened without the support and commitment of Kunz Construction."

❖ ❖ ❖

Building on a long history of federal spending in San Antonio, the United States Air Force begins renovation of a Sony microchip plant at Lackland Air Force Base in 2010 that will become its new cyber warfare intelligence center. One of only four satellite National Security Agency (NSA) cryptologic centers in the United States, NSA Texas will conduct worldwide signals intelligence as well as cyberspace and cybersecurity operations.[18]

"The NSA coming to San Antonio is the equivalent of Toyota coming to San Antonio," John B. Dickson, chairman of the San Antonio Technology Accelerator Initiative, told the *San Antonio Express-News* back in 2007.[19]

The massive cryptologic center complex will occupy fifty acres on Military Drive in Northwest Bexar County.[20] Remediation services company TolTest, as well as Alterman and Whiting-Turner Contracting Company are handling the construction.

This is Alterman's largest single project to date with manpower peaking at 400 electricians. The scope of work is mind-boggling—renovation and conversion of a multi-functional office facility designed to accommodate multiple government organizations, infrastructure for secure information and data systems, uninterruptible power supply systems, auditoriums, and conference rooms to support advanced audiovisual conferencing capabilities.

In 2021, Whiting-Turner will break ground on The University of Texas at San Antonio's $90 million School of Data Science and National Security Collaboration Center.

❖ ❖ ❖

In 2011, Bartlett Cocke L. P. finishes the first building on the new Texas A&M University campus near Toyota, a harbinger for advancement on the city's South Side that now realizes its first four-year higher education offering. The 92,000-square-foot multipurpose space is named for late Senator Frank L. Madla Jr. who, as a member of the Texas legislature, worked with Phil Hardberger and a clowder of civic leaders to secure land for the university.

Its design features local San Saba sandstone, mission concrete tile, and aluminum sun-shaded windows reminiscent of Mission San José.[21] Functionally, the $29 million building houses a reception area, student dining, one-stop Welcome Center for academic offices, a bookstore, instructional and study areas, computer lab, faculty and executive offices, a library, and building facilities service.

Raymond Heath is acting senior project manager, and Kelly Scrimpsher is superintendent. The project receives the 2011 Associated Builders and Contractors Excellence in Construction Merit Award as well as the Excellence in Construction Project of the Year Award from the American Subcontractors Association. More accolades will come for Heath and Scrimpsher during Bartlett Cocke's next chapter of construction on the campus.

Bartlett Cocke and Walbridge Aldinger were part of the winning team that built the game-changing Toyota manufacturing plant.

Main Plaza and downtown before the renovation of the plaza,
City Hall, and City Council Chambers

Chapter Ten
2013–2022

	Central Academic Building, Texas A&M University –San Antonio
	Patriots' Casa, Texas A&M University–San Antonio
Familiar	The DoSeum
& Famous	Tobin Center for the Performing Arts
Places	Temple Beth-El Dome Restoration
	Robert L. B. Tobin Land Bridge
	Oxbow and Credit Human Headquarters
	Navistar Manufacturing Plant
	City Hall Renovation
	Tech Port Center + Arena

In 2013, the *San Antonio Business Journal* digs into the master plan for Texas A&M University–San Antonio. A special supplement features Bartlett Cocke L. P. and the work the firm is performing on Phase II, a $75 million expansion. It's a continuation of the work started with the Senator Frank L. Madla Building.

Enrollment is about 4,000, but the university is planning for 25,000. Key to the expansion is the Central Academic Building, a 170,750-square-foot, H-shaped structure with a grand courtyard for graduations and large gatherings. A 430-seat auditorium is under construction on the building's west side, and "a dramatic four-story arch will rise over the large, embossed, copper-clad doors."[1]

Raymond Heath, senior project manager, and Kelly Scrimpsher, superintendent, are also overseeing the build of the 22,000-square-foot Patriots' Casa, intended to help student veterans and their families transition from the military to academia through career support.[2] Design of the university's "academic home for the brave" includes computer labs, a student veteran's lounge, and family counseling rooms. Eventually, a Survivor Tree, grown from a seedling that survived the September 11 attacks at the Twin Towers, will be planted in a healing garden, its identification plaque fabricated from Ground Zero and Freedom Tower steel.[3]

❖ ❖ ❖

Building on experience with the San Antonio Museum of Art and other high-profile projects, Guido Construction brings to life a new children's museum for the city in 2014. The location of the 104,000-square-foot, two-story DoSeum is that of a former car dealership on Broadway Street, north of downtown.

Producing San Antonio's first new museum in more than fifty years has proven both gratifying and difficult for the builder.

Delighted children and their parents are invited to "break ground" on the museum in a sandbox built just for the occasion. Once construction begins, Project Manager Brad Tatum and Superintendent Tom Sanchez discover contaminated soil on the five-plus-acre site, the culprit a buried 700-gallon ground waste fuel tank. Ordinarily, such a finding would delay a construction project, but through resequencing of tasks as the soil is tested and removed, work carefully continues.

After reshoring the concrete foundation for support, the team pours and erects onsite three two-story, tilt-wall exhibit halls, weighing up to 38,000 pounds and interspersed with glass façades that enclose more than 65,000 square feet of state-of-the-art exhibit space. Crews also transplant two large heritage oak trees and install a bright red serpentine fence made of steel pickets, electric blue glass gabion walls, a water reclamation sculpture, and structures for climbing throughout the ultra-modern, hands-on museum.

Subcontractors contributing to a timely completion within budget include Wilborn Steel Company, O'Haver Plumbing, Big State Electric, Urban Concrete Contractors, and Garden & Ornamental Design. Because of the integrated teamwork, 98 percent of the construction waste is recycled, and soon, the museum will produce 30 percent of its own energy.

❖ ❖ ❖

Another high-profile project happening during this period is the metamorphosis of Municipal Auditorium into the Tobin Center for the Performing Arts–a world-class venue for hosting the city's opera, ballet, and symphony while also attracting international acts. Joint Venture team Linbeck/Zachry renovates and expands the 1926 auditorium, located downtown on the banks of the San Antonio River, with the help of grants, private funds, and venue tax money. Former San Antonio mayor Phil Hardberger is a driving force behind it all and will years later recall, "The construction kept the historic façade of the city's old municipal auditorium and three of its original limestone exterior walls. The rest was dirt and open sky when the contractors began building."

Subcontractor Mike Sireno with BakerTriangle provides drywall and ceiling work, but also "different plasters, acoustical wall panels, and many other specialty items requiring great attention to detail and pre-planning." He too will later reflect, "While the project had its challenges, it is one of those projects that you can be proud to have been a part of and that many people will enjoy for generations to come."

Upon completion in 2014, the state-of-the-art center features a stage house soaring 104 feet, and capacity to seat 2,100. A well-received performance by Paul McCartney kicks off a month-long grand opening celebration. Nearly a decade later, Hardberger will affirm his vision for such a place. "Completed with landscaping and an alternative river barge entrance, it has become an iconic feature of the San Antonio skyline."

Linbeck/Zachry receives the Texas Building Branch-AGC Outstanding Construction Award in 2014.

❖ ❖ ❖

One of the most famous roofs in San Antonio belongs to Temple Beth-El, South Texas' oldest synagogue serving nearly a thousand families and often comprising an all-female clergy.[4] After an April 12, 2016, massive hail swath—that ends up being the second costliest hailstorm in Texas history at $1.3 billion in damages—thousands of clay tiles on the temple's iconic dome are cracked and beyond repair. Beldon Roofing takes on restoring the massive structure, which spans 8,700 square feet, in 2018.[5]

CEO Brad Beldon later recalls, "During our initial evaluation, an obvious challenge was difficult access to the roof and the height and shape of the roof dome."[6] Worker safety is key to the historic renovation, so they build a four-tier scaffolding system for approaching the dome's base, which is seventy feet off the ground.[7]

Since blueprints are nonexistent for the ninety-one-year-old temple, the company turns to technology to guide them. The crew flies a drone fitted with a high-definition camera over the roof to capture images and record video. Then, using a digital laser, measurements are recorded, and a long-hidden geometric pattern is discovered.

Workers remove the damaged tiles, battens, and underlayment by hand. Debris is carefully lowered by crane to the ground as precautions are taken not to interfere with religious services and the on-campus elementary school's activities.[8]

Over the span of four months, 33,000 new tiles are installed along with a copper cap, and luster and grandeur is restored to the temple on the hill. Beldon receives a preservation award from The Conversation Society of San Antonio and is named a Gold Circle Awards finalist for outstanding workmanship by the Roofing Alliance.[9]

❖ ❖ ❖

With great hurrah, SpawGlass completes the 150-foot-wide Robert L. B. Tobin Land Bridge in late 2020 as the largest dedicated wildlife overcrossing in the United States.[10] Early the following year, an ADA-compliant skywalk opens alongside the bridge guiding visitors over treetops to the landscaped pinnacle.

But there is history.

In the lead-up to a 2017 city bond election, the groundwork is laid for a unique land bridge and wildlife crossing at Phil Hardberger Park. The park's master plan calls for the bridge to be built over six-lane Wurzbach Parkway, connecting two sections of the popular 311-acre natural area.

Critics dismiss the projected $25 million project as a "critter bridge" and a waste of taxpayer money. But the editorial staff at the *San Antonio Express-News* opines in favor, writing, "There should be no hesitation about putting the Hardberger land bridge on the city bond."[11] AGC's Doug McMurry, a veteran of the Main Plaza controversy and dozens of bond elections, weighs in with an op-ed piece, "Whether it is building a figurative bridge between divided communities or constructing a real one to unite Phil Hardberger Park, it is a task worth doing."

The park's well-connected, not-for-profit conservancy board of directors agrees to raise $10 million in private funds. So, if the voters approve $15 million, taxpayers will get a $25 million, first-of-its-kind land bridge for $15 million. It seems like a no-brainer.

Yet, opposition persists until a city parks bond committee snuffs out detractors with a compromise and moves a $13 million recommendation along to the city council. With the support of a dozen community and trade groups, including AGC and the local chapter of the American Institute of Architects, voters approve $850 million in bond funding, including $13 million for the bridge.

❖ ❖ ❖

A year never to be forgotten, 2020 is also when COVID-19 strikes, sending shockwaves around the globe from Wuhan, China to Waxahachie, Texas. Eventually, the pandemic will claim more than 6.4 million lives worldwide, including a million-plus in the United States.

In March, as the reality of the highly contagious virus sinks in, contractors from around the country are scrambling to leave Las Vegas, the site of the annual convention of the AGC of America. Mayors and governors are issuing emergency orders closing businesses and prohibiting large gatherings.

In San Antonio, thousands of employees from USAA, Valero, and SWBC begin working from home.[12] Conventions and sporting events are canceled. Members of the local chapter of the AGC are concerned about worker safety and their contractual obligations with owners and clients as infections sweep the city.

Local, state, and national AGC staff quickly coordinate a massive effort to ensure strict safety protocols and guarantee the construction industry's "essential" status. The San Antonio chapter is in close communication with its members and with San Antonio Mayor Ron Nirenberg, promoting safety and a "Together, we'll get through this" campaign.

Overcoming these enormous challenges exacerbated by supply chain interruptions, Joeris General Contractors is able to complete a pair of projects on three acres of lower Broadway Street across from historic Pearl. The Broadway Office Development includes the eight-story Credit Human headquarters building and the four-story multi-tenant Oxbow building, both built over structured parking. Like the University of the

Incarnate Word skyline projects, these buildings are highly visible from US Highway 281. Sustainability measures, including rainwater capture and geothermal wells for heating and cooling, earn Joeris the San Antonio Business Journal Best Office Development and Best Green Project awards, a Golden Trowel Award from the San Antonio Masonry Contractors Association, and the Texas Building Branch-AGC Outstanding Construction Award.

❖   ❖   ❖

After being jilted by the San Antonio Spurs on the SBC Center a decade earlier, Turner Construction wins big with Navistar's San Antonio Manufacturing Plant in 2022. The $250 million project is further evidence that the manufacturing Toyota brought to town is expanding. Mayor Nirenberg sums it up at the ribbon cutting ceremony, "This is more proof that San Antonio's advanced manufacturing sector is booming." [13]

With the help of $21 million in incentives that include tax breaks, grants and fee waivers, the South Side plant puts 500 mostly local residents to work, aiming for 650, to produce both diesel and electric trucks and construction vehicles through sustainable practices and revolutionary 4.0 technologies. [14] The five-building manufacturing facility totaling nearly one million square feet includes a general assembly area, body and paint shops, a logistics center, and a truck specialty center.

Because the construction began during the pandemic, Turner enforced a single access point to enter the site and required temperature monitoring stations. The company also mandated proper face masking, dispersed break areas, and staggered start times to reduce worker congestion.

Subcontractors Steel Masters, Patriot Erectors, Big State Electric, and Johnson Controls played major roles during construction of the assembly plant.

Ultimately, full production will reach fifty-two vehicles a day and serve as a benchmark for Navistar's national manufacturing network.[15]

❖　❖　❖

Another general contractor facing COVID challenges on a high-profile project is Guido Construction. The renovation of San Antonio's historic City Hall is fraught with worker shortages, supply and delivery disruptions, and constant fear of contagious infections.

Project Executive Albert Gutierrez, Project Manager Ashley Ruggles, and Superintendent Lonnie Voigt guide the $40 million project, peeling back layers of additions and remodeling since its original construction in 1892. Their goal is to bring the building into the modern age with updated fire protection, advanced security systems, and green building elements such as solar hot water, electric vehicle charging stations, and the latest in communications technology.

The company employs artisans and subcontractors with experience restoring historic features such as exterior limestone, cast stone façades, and interior plaster wall and ceiling finishes, and they work painstakingly for three years. Among the subcontractors are Central Electric, Curtis Hunt Restorations, and Phoenix I Restoration and Construction.

Crews build a new ADA-accessible east entrance and uncover fourteen-foot windows that allow for natural light and sweeping views. They also create an intricate ventilation system to protect the roots of historic pecan and ash trees as the work on the building picks up pace.

As with most city projects there are requirements to work with a certain percentage of small, minority, and women-owned businesses. Despite pandemic-related challenges, Guido exceeds the requirements and goes on to win nine prestigious awards for the renovation. Among the awards are the San Antonio Business Journal Building San Antonio Award, The Conservation Society of San Antonio Historic Preservation Building Award, and the AGC of America's Build America Award for Historic Preservation.

❖ ❖ ❖

In 2022, Sundt Construction completes one of the most talk-ed-about buildings in recent San Antonio history, the $53 million, 130,000-square-foot Tech Port Center + Arena, built to host concerts, e-sports tournaments, product launches and large training programs. It required 62,000 tons of aggregate base to construct the building's 109,000-square-foot pad, which crews worked double shifts to complete in only a week.

A multipurpose arena—one component of the mega technology and innovation campus—is somewhat reminiscent of a time when the Convention Center Arena was the go-to entertainment spot. Except, in the twenty-first century, it's not a venue for Rod Stewart, Led Zeppelin, and Sonny & Cher, but rather the likes of Boyz II Men, Rise Against, and the Smashing Pumpkins, the with-it arena's inaugural act.

The campus is also the new home for the San Antonio Museum of Science and Technology, a showroom to demonstrate technological innovations as well as a 24-hour electronic gaming area, full-scale food hall, and maker space laboratory.

Bob Aniol is the superintendent. Subcontractors F. A. McCo-mas, TD Industries, Alpha Waterproofing, and Quality Fence & Welding help him make it all happen.

Tech Port Center + Arena is located on the city's southwest side, on the 1,900-acre Port San Antonio campus, home to more than eighty employers and their more than 14,000 employees that include leading names in aerospace, cybersecurity, robotics, defense, and other advanced operations. The *San Antonio Business Journal* is looking ahead to new development at the port that will attract more companies, talent, and visitors, with Tech Port Center + Arena leading the "next-gen pivot."[16]

*Constructed by Bartlett Cocke L. P., Texas A&M University–
San Antonio's Patriots' Casa helps veterans and their families
transition from the military to academia.*

*Joeris General Contractors completed
the Oxbow and Credit Human Headquarters projects
on lower Broadway, across from the old Pearl Brewery.*

Cosmo F. Guido (seated center with family at the DoSeum) was the son of an Italian immigrant and a driving force behind the company's success.

The Linbeck/Zachry joint venture renovated and expanded the
Tobin Center for the Performing Arts.

SpawGlass finished the first-of-its-kind Robert L. B. Tobin Land Bridge in late 2020.

*Guido Construction earned multiple awards for its renovation
of San Antonio's historic City Hall.*

Tech Port Center + Arena, built by Sundt Construction, is located on the 1,900-acre Port San Antonio campus.

Turner Construction built the $250 million Navistar Truck Assembly Plant.

Chapter Eleven
2023

This is the year the local chapter of AGC celebrates its one hundred years of building San Antonio. While many architects, engineers, and contractors know who helped build the nation's seventh largest city, the vast majority of the 1.4 million people who live in San Antonio do not. Nor do the approximately forty million who visit the Alamo City annually.

Just as knowing something about the chef who prepared the dish almost always makes the meal more enjoyable, knowing a little about who built the familiar and famous places provides meaning and context to a city.

In San Antonio, construction companies change and adjust to

meet the needs of a city with an appetite for growth. There is a sense of rhythm and purpose that begins in the past but leans into to the future. Oftentimes, new construction happens on the fringes while older parts experience revitalization.

And sometimes individual building uses themselves change. For example, the historic Tower Life Building, as reported by the *San Antonio Business Journal* in August 2022, could be transformed into new residential space.[1] What was once the celebrated skyscraper with offices is now reimagined as a solution to an identified housing shortage.

❖ ❖ ❖

Since its purchase by developer and former contractor Mitch Meyer in 2022, the old Builders Exchange Building is undergoing cosmetic upgrades. The former office building is listed on the National Register of Historic Places. Renovated in 1993 to include apartments, Meyer is making additional improvements so downtown renters may enjoy modern amenities.[2] Upgrades to the space once occupied by directors of the AGC will likely please preservationists at The Conservation Society of San Antonio.

❖ ❖ ❖

In 2023, high-rise living is also the focus of Rogers-O'Brien Construction. The company is constructing 300 Main, a thirty-two-story building downtown with 354 multifamily units, more than 6,000 square feet of retail space, a sky lounge, and six levels of above-grade structured parking.[3] Troy Wylam is the senior superintendent and is working with Haley-Greer, Kilgore Industries, and Steel Designs. Like most superintendents under pressure, Wylam is managing deadlines, labor shortages, and supply chain disruptions.

❖ ❖ ❖

On the East Side, affordable housing is again part of the evolving construction story. In 2022, the US Department of Housing and Urban Development conditionally approved financing for the renovation of an abandoned Friedrich Air Conditioning factory building into mixed-rate apartments known as Friedrich Lofts.[4] Ultimately, the $80 million project is expected to provide 358 units of varying prices.

❖　❖　❖

Nearby, the multi-phase Echo East development is much further along in the construction process.[5] Like several other projects in San Antonio, public financing is a key ingredient. Construction financing includes $750,000 from Bexar County, $425,000 from the George Gervin Youth Center, $2.5 million from San Antonio's inner city Tax Increment Reinvestment Zone, and approximately $18 million in tax credit equity from the state of Texas.

State Representative Barbara Gervin-Hawkins and the George Gervin Youth Center, which was founded by Hawkin's brother, San Antonio Spurs legend George Gervin, are building a mixed-use "village" on approximately twenty acres near the AT&T Center. Just as Kit Goldsbury and Silver Ventures imagined a hotel, housing, and restaurants revitalizing Pearl Brewery, Gervin-Hawkins dreams of a comparable transformation on the East Side.

"The community has wanted something so bad," she tells the *San Antonio Business Journal.* "I think doing not only just housing but bringing commercial there for services as well as job creation is a critical aspect."[6]

Turner Construction is helping to make that dream a reality by constructing the first phase of the $130 million development, 192 apartments.

❖　❖　❖

Across town on the near West Side, former San Antonio AGC chapter president Rene Garcia is involved in redeveloping the Scobey industrial complex owned by VIA Metropolitan Transit.[7] As a co-founder of the DreamOn Group—a privately held firm that aids in transforming communities—Garcia is working to build 81,000 square feet of combined affordable and market-rate housing, 32,000 square feet of retail and education space, 26,500 square feet of offices, and 27,500 square feet of storage.[8]

Garcia and partner Julissa Carielo are pursuing state and federal historic tax credits, which would cover about a third of the construction costs. They are also seeking a property tax waiver for the residential portion and planning a long-term lease agreement with VIA for the office space.[9]

❖ ❖ ❖

On the city's far West Side, multifamily general contractor Galaxy Builders is completing 384 apartments at Kallison Ranch. Superintendent Fred Summers is working on the $42 million project with Ranger Excavating, Cappadonna Electric Contractor, and Central Texas Lath & Plaster. The team faces challenges of volatile material prices, the availability of materials in general, and not interfering with adjacent projects going on simultaneously. Mata Drywall, Sonny's Plumbing, CWT Construction, and Southern Framing Services are also important contributors on the site.

❖ ❖ ❖

While multifamily housing and mixed-use development targeting locals is abundant in 2023, there is something for visitors, too. Guido Construction is renovating the historic Crockett and Woolworth buildings across from the Alamo. The result will be a new $25 million Alamo Visitor Center and Museum with 100,000 square feet of event space, a rooftop restaurant, and museum galleries, slated for a 2025 opening.[10] The center will also house British musician Phil Collins' entire Texana Collection, consisting of hundreds of Alamo and Texas Revolution artifacts including a rifle and leather pouch owned by Davy Crocket, and a knife that belonged to Jim Bowie.[11]

❖　❖　❖

Museums, Olmos Dam, the River Walk, and San Antonio International Airport are part of the city's rich and vibrant history. So are hospitals, universities, military bases, and convention hotels. That history is more complete when the builders and renovators of the these, and many other places where millions work, play, shop, study, and worship are included. Like the determined individuals at the Alamo, they belong in the San Antonio story.

Turner Construction helps make the Echo East dream a reality.

Acknowledgments

The authors thank the following individuals for their help and support. Without their contributions, this book would not have been possible.

Richard Alterman

Blaine Beckman

Mike Boyle

Angela Cardwell

Mike Cooney

Andi Galloway

Lauren Guido Tew

Kirk Kistner

Andy Mitchell

Lane Mitchell

Doug Nunnelly

Diane Prodger

Mick Prodger

Beth Standifird

John Wright

Notes

Chapter One

[1] "New Yankee Baseball Stadium Topped By Copper Cornice Weighing 15 Tons," *The Olean Evening Herald*, January 16, 1923, 10.

[2] "History," Alterman, accessed August 27, 2022, https://goalterman.com/alterman/history/.

[3] Paula Allen, "Medical Arts Building now landmark hotel," *San Antonio Express-News*, March 7, 2015, https://www.expressnews.com/150years/education-health/article/Medical-Arts-Building-served-doctors-patients-6119325.php.

[4] "1925 To Be San Antonio's Great Building Year," *San Antonio Express*, January 1, 1925, Financial Section.

[5] "S.A. Leads South With Skyscraper," *San Antonio Light*, Real Estate and Building, May 31, 1925, 1.

[6] "Bulk of City's $5,000,000 Bond Issue To Be Spent On Improvements in 1925," *San Antonio Express*, January 1, 1925, 2A.

[7] "Olmos Dam," Texas Beyond History, accessed June 13, 2022, https://www.texasbeyondhistory.net/st-plains/images/ap10.html.

[8] "The Milam Building," Weston Urban, accessed August 27, 2022, https://westonurban.com/project/the-milam-building/.

[9] Paula Allen, *San Antonio: Then and Now*, (United Kingdom: Pavilion Books, 2015), 48.

[10] The San Antonio Chapter of the American Institute of Architects, *San Antonio Architecture: Traditions and Visions*, (San Antonio, Texas: AIA San Antonio 2007), 216.

[11] "McNay Moments: From Mansion to Museum: The Metamorphosis of Mrs. McNay's Home," The McNay Art Museum, February 21, 2014, https://www.mcnayart.org/blog/mcnay-moments-from-mansion-to-museum-the-metamorphosis-of-mrs-mcnays-home/.

[12] Benjamin Olivo, "Paula Allen: 31-story Smith-Young building opened in '29," *San Antonio Express-News*, June 20, 2010, https://www.mysanantonio.com/life/life_columnists/paula_allen/article/Paula-Allen-31-story-Smith-Young-building-opened-783158.php.

[13] National Register nomination, Smith Young Tower, prepared by Stephanie Hetos Cocke, 1991.

[14] Godfrey Lane, "Willard Eastman Simpson, P.E. 1883–1967", January 4, 2017, 10, https://silo.tips/download/willard-eastman-simpson-pe.

[15] Benjamin Olivo, "Express-News building marked end of boom," *San Antonio Express-News*, https://www.expressnews.com/150years/economy-business/article/Express-News-building-marked-end-of-boom-6184580.php.

[16] Olivo, "Express-News building marked end of boom."

[17] "Our History," Guido Companies, accessed August 27, 2022, https://www.guidoco.com/why-guido.

[18] Dorothy Steinbomer Kendall, "San Pedro Springs Park," Texas State Historical Association, updated May 15, 2018, https://www.tshaonline.org/handbook/entries/san-pedro-springs-park.

[19] "School History," San Antonio Independent School District, accessed August 27, 2022, https://schools.saisd.net/page/007.history.

[20] *Meeting Minutes 1924-1936,* San Antonio Chapter of the Associated General Contractors.

Chapter Two

[1] "Hipolito F. Garcia Federal Building and U.S. Courthouse, San Antonio, TX," US General Services Administration, accessed March 9, 2022, https://www.gsa.gov/historic-buildings/hipolito-f-garcia-federal-building-and-us-courthouse-san-antonio-tx.

[2] "New $2,225,000 S. A. Post Office Open Monday," *The San Antonio Light,* October 10, 1937,13; "Hipolito F. Garcia Federal Building and U.S. Courthouse, San Antonio, TX," US General Services Administration, accessed March 9, 2022, https://www.gsa.gov/historic-buildings/hipolito-f-garcia-federal-building-and-us-courthouse-san-antonio-tx; Patrick Danner, "S.A. federal building is a 'national treasure'," *San Antonio Express-News*, December 17, 2017, https://www.expressnews.com/150years/economy-business/article/Downtown-federal-building-a-national-6272071.php.

[3] "Hospital Seeks Another Million," *San Antonio Express,* November 13, 1936, 7.

[4] "About Us," Brooke Army Medical Center, March 4, 2022, https://bamc.tricare.mil/About-Us.

[5] Sig Christenson, "A hospital born in humble roots becomes a giant," *San Antonio Express-News,* January 31, 2015, https://www.expressnews.com/150years/education-health/article/A-hospital-born-in-humble-roots-becomes-a-giant-6052269.php.

[6] "Predict 18,000 Crowd At Stadium," *The San Antonio Light,* December 27, 1939, 7-A.

[7] Richard Hurd, "San Antonio River Walk," accessed August 29, 2022, https://aggie-horticulture.tamu.edu/plantanswers/riverwalk/Pages/slide26.html; "River Walk–San Antonio TX," The Living New Deal, https://livingnewdeal.org/projects/river-walk-san-antonio-tx/.

[8] "Robert Hugman," Oral History Program, Bexar County Historical Commission, February 17, 1977, https://digital.utsa.edu/digital/collection/p15125coll4/id/1295.

Chapter Three

[1] Ed. Claudia R. Guerra, *300 Years of San Antonio and Bexar County,* (San Antonio, Texas: Trinity University Press, 2019), 239.

[2] "Soaring to New Heights," Flashback Friday, G.W. Mitchell Construction, 2021.

[3] "Lee Christy Renamed Head of Builders," *San Antonio Express,* July 4, 1943, 10.

[4] The San Antonio Chapter of the American Institute of Architects, *San Antonio Architecture: Traditions and Visions,* (San Antonio, Texas: AIA San Antonio, 2007),189.

[5] René A. Guzman, "Nogalitos H-E-B opened in 1945 as 'store of tomorrow.' 76 years later, the South Side grocery store still ahead of its time," *San Antonio Express-News,* August 11, 2021, E1.

[6] "Alameda Theater Set for Opening Night," *San Antonio Light,* March 9, 1949, 1-D.

[7] Bartlett Cocke Jr., *Bartlett Cocke General Contractors: The First 50 Years, 1959 to 2009,* (San Antonio, Texas: The Watercress Press, 2009), 57.

[8] "Alamo Hts. Contract is $737,625," *San Antonio Light,* May 8, 1949, Section E.

[9] "Sunset Ridge Stores Open for Business," *San Antonio Light,* June 29, 1951, 1.

[10] "Kelly Buildings Let at $923,873," *San Antonio Express,* July 1, 1951, p.3.

[11] The San Antonio Chapter of the American Institute of Architects, *San Antonio Architecture: Traditions and Visions,* 201.

[12] "Ahead of Schedule," *San Antonio Express,* January 13, 1952, 3.

Chapter Four

[1] Brian Purcell, "San Antonio Area Roads History, Freeway System History," The Texas Highway Man, updated December 5, 2021, https://www.texashighwayman.com/fwy-history.shtml.

[2] "S.A. Area Has Many Members in AGC Chapter," *San Antonio Express,* February 22, 1953, 4.

[3] "8-Week S.A. Building Strike Comes to End," *San Antonio Express,* September 21, 1953, 1.

[4] "Workers on S.A. Construction Projects Strike," *San Antonio Light,* August 3, 1953, 21.

[5] The San Antonio Chapter of the American Institute of Architects, *San Antonio Architecture: Traditions and Visions* (San Antonio, Texas: AIA San Antonio, 2007),199.

[6] Nancy Haston Foster and Benjamin A. Fairbanks, *The Texas Monthly Guidebooks, San Antonio, Revised Fourth Edition,* (Houston, Texas: Gulf Publishing Company, 1994), 35.

[7] The San Antonio Chapter of the American Institute of Architects, *San Antonio Architecture: Traditions and Visions,* 223.

[8] "Campus Facilities Greatly Improved," *The San Antonio Light,* August 3, 1954, 21.

[9] "College Building Bidders Named," *San Antonio Express,* August 12, 1953, 2.

[10] "St. Philip's to be In New Building," *San Antonio Express,* August 3, 1954, 1-C.

[11] Ed. Kathryn O'Rourke, "1959 History and Development of La Villita Assembly Hall," *O'Neil Ford on Architecture,* (New York, USA: University of Texas Press, 2021), 99, https://www.degruyter.com/document/doi/10.7560/316382-010/html?lang=en; Godfrey Lane, "Willard Eastman Simpson, P.E. 1883-1967," January 4, 2017, 12, https://silo.tips/download/willard-eastman-simpson-pe.

[12] Madison Iszler, "CPS Energy finds buyer for historic Villita Assembly Building," *San Antonio Express-News,* March 2, 2020, https://www.expressnews.com/real-estate/article/CPS-Energy-finds-buyer-for-historic-Villita-15099959.php#:~:text=March%202%2C%202020%20Updated%3A%20March,cost%20of%20its%20new%20headquarters.

[13] Brian Purcell, "San Antonio Area Roads History, Freeway System History," The Texas Highway Man, updated December 5, 2021, https://www.texashighwayman.com/fwy-history.shtml.

[14] Richard A. Feinberg and Jennifer Meoli, "A Brief History of the Mall," in *NA-Advances in Consumer Research Volume 18,* eds. Rebecca H. Holman and Michael R. Solomon (Provo, Utah: Association for Consumer Research,1991), 426-427.

[15] "60 Acres of Shopping Delight," *San Antonio Express and News,* September 25, 1960, 2.

[16] "Work on South Texas' Biggest Center Begun," *San Antonio Express and News,* January 9, 1960,1-B.

[17] Advertisement, *San Antonio Express and News,* Sept. 23, 1960, 19.

[18] Joan Didion, "On the Mall," *The White Album,* (New York: Ferrar, Straus and Gireaux, 1979), 180,186.

[19] "60 Acres of Shopping Delight," 2.

[20] "Work on South Texas' Biggest Center Begun," 1-B.

[21] "That's North Star Mall," *San Antonio Express and News,* September. 25, 1960, 2.

[22] "McCreless Center Takes Shape," *San Antonio Light,* August 27, 1961, 5-D.

[23] SnappyBob, "McCreless Mall," City-Data.com forum, accessed July 9, 2022, https://www.city-data.com/forum/san-antonio/2150120-mccreless-mall.html.

[24] "Sommers Store In New Center," *San Antonio Express,* March 29, 1962, 6; "Sommers Will Broil Your Meal," *San Antonio Express,* March 29, 1962, 6.

[25] "Newest Store in Big M City," *San Antonio Express*, April 12, 1962, 2-C; "Host of Specials Ready for Opening," *San Antonio Express*, April 12, 1962, 2-C.

[26] "Penney's 60th Anniversary: Your New Penney's opens Thursday, March 29 in McCreless Shopping City," *San Antonio Express,* March 29, 1962, 4.

[27] Advertisement, *San Antonio Light,* March 28, 1962, 18-B.

Chapter Five

[1] "Lights Give Center Float Effect," *San Antonio Light,* May 24, 1963, 39; "New Volkswagen Center Will Open Today," *San Antonio Express,* May 24,1963, 2-B.

[2] "Browning General Contractor for Job," *San Antonio Express and News,* August 15, 4-J.

[3] "HemisFair Model Is on Display," *San Antonio Express and News,* August 15, 1965, 2-J.

[4] "Convention Center Abuilding," *San Antonio Express,* March 23, 1966, 11-A; Ed. Claudia Guerra, *300 Years of San Antonio and Bexar County,* (San Antonio, Texas: Trinity University Press, 2019), 21.

[5] "Convention Center Abuilding," *San Antonio Express,* March 23, 1966, 11-A; "The Most Exciting 92 Acres in Southwest," *San Antonio Light,* January 1, 1967, 1.

[6] "S.A. Creating New Social, Civic Hub," *San Antonio Light,* October 30, 1966, 69-NC.

[7] "New Bids Examined," *San Antonio Light,* August 3, 1966, 67.

[8] Frank Duane, Texas State Historical Association Handbook of Texas, "HemisFair '68," https://www.tshaonline.org/handbook/entries/hemisfair-68.

[9] Frank Duane, "HemisFair '68," Handbook of Texas, Texas State Historical Association, published 1976, updated March 30, 2021, https://www.tshaonline.org/handbook/entries/hemisfair-68.

[10] "Host of Officials Breaks Ground for Fair Pavilion," *San Antonio Express,* February 21, 1967, 1.

[11] "New Labor Fuss Perils Fair Work," *San Antonio Express,* May 19, 1967, 16-A.

[12] Bartlett Cocke Jr., *Bartlett Cocke General Contractors: The First 50 Years, 1959 to 2009,* (San Antonio, Texas: The WaterCress Press, 2009), 9-13.

[13] "Pavilion Construction to Begin," *San Antonio Light,* February 12, 1967, 10-C.

[14] "The Most Exciting 92 Acres in the Southwest," *San Antonio Light,* January 1, 1967, D-1.

[15] "San Antonio, United States 1968, Hemisfair '68," America's Best History, accessed August 30, 2022, https://americasbesthistory.com/wfsanantonio1968.html.

[16] Madison Iszler, "San Antonio firm getting another shot at Hemsfair development with new deal from City Council," *San Antonio Express-News,* April 7, 2022, https://www.expressnews.com/business/article/Hemisfair-project-Zachry-Hospitality-17065035.php.

[17] Zachry Construction Corporation, accessed August 30, 2022, https://www.zachryconstructioncorp.com/Projects/Building-Historical/Hilton-Palacio-del-Rio-Renovations/; "Hilton Gets Briefing On Hotel," *San Antonio Express,* August 29, 1967, 4-D.

[18] "South Texas Medical School: Transforming San Antonio's Medical Community," Flashback Friday, G.W. Mitchell Construction, 2021; "UT Health San Antonio's History," UT Health San Antonio, accessed August 30, 2022, https://uthscsa.edu/university/history.

[19] "Lieutenant Governor tours Medical Construction Site," *San Antonio Express,* November 22, 1966, 8-D.

[20] Doris Wright, "Trinity's Laurie Auditorium Dedicated," *San Antonio Light,* October 31, 1971, 2-A.

[21] "Trinity Previews Laurie Auditorium," *San Antonio Express,* October 21, 1971, 2-C.

[22] "Browning Company Builds Army's New Med School," *San Antonio Express and News,* December 9, 1972, 7-D.

Chapter Six

[1] "Handy Andy is U.S. Top Food Retailer," *San Antonio Express,* May 11, 1973, 9-F.

[2] "Murphy Hospital Draws Acclaim," S*an Antonio Light,* November 18, 1973, 2-A; "New VA Hospital to Open," *San Antonio Express and News,* October 14, 1973, 1; "2,000 Tour Massive New Audie Murphy Hospital," *San Antonio Light,* November 18, 1973, 18-A.

[3] Advertisement, *San Antonio Light,* November 17, 1973, 7-D.

[4] "Bank Tower For Frost All Ready," *San Antonio Express and News,* September 2, 1973, 15-D; "Gold Key Opens Doors to Frost," *San Antonio Express and News,* September 8, 1973, 3-A.

[5] "Bank Tower For Frost All Ready," 15-D.

[6] "Construction Problems Delay Full Enrollment at UTSA," *San Antonio Light,* February 2, 1974, 9-A.

[7] "UTSA Construction Suit Filed," *San Antonio Light,* August 5, 1976, 8-A.

[8] "Government must reduce spending, says economist," *San Antonio Express,* November 8, 1974, 3-C.

[9] "S.A. municipal building nears $165 million mark," *San Antonio Express-News,* February 2, 1975, 2-C.

[10] "Economy Surges Upward," *San Antonio Light,* October 20, 1975, 1.

[11] "New Complex Brings USAA Together," *San Antonio Light,* May 2, 1976, 8-D.

[12] Bartlett Cocke Jr., *Bartlett Cocke General Contractors: The First 50 Years, 1959 to 2009,* (San Antonio, Texas: Watercress Press, 2009),16-17.

[13] "San Fernando: Restoring the past," *San Antonio Express-News,* December 14, 1975, 4-B; "Downtown Building," *San Antonio Express-News,* Action Express, 5-A.

[14] "Raising the Roof on the Hemisfair Arena," Flashback Friday, G.W. Mitchell Construction, 2021.

[15] "Riverwalk Marriott, San Antonio," *Texas Architect,* January/February 1980.

[16] "Big-city flair for downtown," *San Antonio Express-News,* December 27, 1981, 4-M.

[17] Cocke Jr., *Bartlett Cocke General Contractors: The First 50 Years, 1959 to 2009,* 28-30.

[18] Tricia Lynn Silva, "USAA Buys One Riverwalk Place," *San Antonio Business Journal,* August 29, 2013, https://www.bizjournals.com/sanantonio/blog/2013/08/usaa-buys-one-riverwalk-place.html#:~:text=USAA%20Real%20Estate%20Co.%20is,owner%20of%20One%20Riverwalk%20Place.&text=A%20well%2Drecognized%20property%20in,district%20has%20a%20new%20owner.

Chapter Seven

[1] John C. Ferguson, "Interfirst Plaza," *Texas Architect,* September/October 1983, 48.

[2] "USAA buys Bank of America Plaza," *San Antonio Business Journal,* August 16, 2017, https://www.bizjournals.com/sanantonio/news/2017/08/16/usaa-buys-bank-of-america-plaza.html; Robert Rivard, "San Antonio's First $100 Million Tower Sale Looming," *San Antonio Report,* November 1, 2014, https://sanantonioreport.org/san-antonios-first-100-million-tower-sale/.

[3] The San Antonio Chapter of the American Institute of Architects, *San Antonio Architecture: Traditions and Visions,* (San Antonio, Texas: AIA San Antonio, 2007), 74.

[4] "Pace Foods Headquarters," *Texas Architect,* September/October,1983, 54–55.

[5] Jan Jarboe Russell, "What Does H.E.B. Stand For, Anyway?" *Texas Monthly,* April 1988, 26.

[6] Bartlett Cocke Jr., *Bartlett Cocke General Contractors: The First 50 Year, 1959 to 2009,* (San Antonio, Texas: The WaterCress Press, 2009), 38.

[7] Tim Griffin, "Palo Alto chief watches school birth," *San Antonio Express-News,* May 26, 1986.

[8] Kunz Construction Company, "Hemisfair Plaza Redevelopment," 1988.

[9] Ismael Perez, "San Antonio's Rivercenter Mall opened 31 years ago this month,"

San Antonio Express-News, Feb. 20, 2019. https://www.mysanantonio.com/entertain-ment/article/San-Antonio-s-Rivercenter-Mall-opened-31-years-13631883.php#:~:tex-t=Feb.,20%2C%202019%20Updated%3A%20Feb.

[10] Rebecca Salinas, "A look back at the painstaking half-mile journey of the 1,600-ton Fairmount Hotel in San Antonio," *San Antonio Express-News,* Feb. 24, 2015.

[11] "Downtown Renaissance Looks to Retail 'Eden' on the River Walk," Texas Architect, September/October 1988, 9.

[12] Ismael Perez, "San Antonio's Rivercenter Mall opened 31 years ago this month," *San Antonio Express-News,* February 20, 2019.

[13] Ibid.

[14] Diana J. Kleiner, "H-E-B," Texas State Historical Association, November 1, 1995.

[15] "History," Manhattan Construction Group, https://manhattanconstructiongroup.com/#history.

Chapter Eight

[1] Texas State Historical Association, "San Antonio: 300 Years of History," *Texas State Historical Association Handbook of Texas Online,* (Austin, Texas: Texas State Historical Association, 2020), 56; "Military Bases: Analysis of DOD's Recommendations and Selection Process of Closures and Realignments," US Government Accountability Office, April 15, 1993.

[2] The San Antonio Chapter of the American Institute of Architects, *San Antonio Architecture: Traditions and Visions,* (San Antonio, Texas: AIA San Antonio, 2007), 146.

[3] Nelson W. Wolff, *Transforming San Antonio: An Insider's View of the AT&T Center, Toyota, the PGA Village, and the River Walk Extension,* (San Antonio, Texas: Trinity University Press, 2008), 18.

[4] "Alamodome Opens with the Help of Many AGC Firms," *Construction Leader,* 1993, 9.

[5] Bartlett Cocke Jr., *Bartlett Cocke General Contractors: The First 50 Year, 1959 to 2009,* (San Antonio, Texas: The Watercress Press, 2009), 47.

[6] Legoretta, Projects, San Antonio Public Library, accessed August 31, 2022, https://www.legorreta.mx/en/proyecto-biblioteca-central-de-san-antonio.

[7] "Library In Texas: A Shade Too Much?" *The New York Times,* November 26, 1995, Section 1, 24.

[8] Madalyn Mendoza, "San Antonio's iconic 'Enchilada Library gets a fresh look," *San Antonio Express-News,* July 14, 2021, https://www.mysanantonio.com/news/local/article/san-antonio-enchilada-library-downtown-repaint-16313976.php.

[9] "San Antonio City Council Chambers, Municipal Plaza," City of San Antonio Department of Arts & Culture, accessed August 31, 2022, https://events.getcreativesanantonio.com/venue/san-antonio-city-council-chambers-municipal-plaza/.

[10] Dorothy Steinbomer Kendall, "San Pedro Springs Park," Texas State Historical Association, published September 1, 1995, updated May 15, 2018, https://www.tshaonline.org/handbook/entries/san-pedro-springs-park.

[11] Doug McMurry, "Enormous opportunities created," *San Antonio Construction News* Feature Publication Supplement, November 2002.

[12] Wolff, *Transforming San Antonio: An Insider's View of the AT&T Center, Toyota, the PGA Village, and the River Walk Extension,* 45.

[13] W. Scott Bailey, "Arena project scores a contracting victory," *San Antonio Business Journal,* Week of March 1-7, 2002, 63.

[14] McMurry, "Enormous opportunities created."

[15] Bailey, "Arena project scores a contracting victory," 63.

Chapter Nine

[1] "Toyota sites new plant in Texas for some unusual reasons," *Engineering News-Record,* Feb. 17, 2003, 13.

[2] Nelson W. Wolff, *Transforming San Antonio: An Insider's View of the AT&T Center, Toyota, The PGA Village, and The River Walk Extension,* (San Antonio, Texas: Trinity University Press, 2012), 88.

[3] Ibid, 116.

[4] Greg Jefferson, "Toyota plant stirs labor pot," *San Antonio Express-News,* May 12, 2003, 9A; W. Scott Bailey, "Toyota's plant may be built using union labor, *San Antonio Business Journal,* March 2, 2003, https://www.bizjournals.com/sanantonio/stories/2003/03/03/story2.html.

[5] Barbara Powell, "Foundation building," *San Antonio Express-News,* July 29, 2004, 1E.

[6] Jefferson, "Toyota plant stirs labor pot," 9A.

[7] Ibid.

[8] Powell, "Foundation building," 1E.

[9] Randy Diamond, "Toyota investing $391 million in San Antonio plant," *San Antonio Express-News,* September 17, 2019, https://www.mysanantonio.com/business/local/article/Toyota-investing-391-million-in-San-Antonio-plant-14446084.php.

[10] Bartlett Cocke Jr., *Bartlett Cocke General Contractors: The First 50 Years, 1959-2009,* (San Antonio, Texas: The Watercress Press, 2009), 55.

[11] James Aldridge, "Health Science Center lands $25 million gift for children's health," *San Antonio Business Journal,* January 18, 2007, https://www.bizjournals.com/sanantonio/stories/2007/01/15/daily33.html.

[12] "City of San Antonio Animal Care Services," City of San Antonio, accessed August 31, 2022, https://www.sanantonio.gov/Animal-Care/Home.

[13] Vincent T. Davis, "San Antonio reaches no-kill goal, but ACS has sights set even higher," *San Antonio Express-News,* August 31, 2015, https://www.expressnews.com/news/local/article/San-Antonio-on-cusp-of-reaching-no-kill-goal-6476423.php.

[14] "About," The Chapel of the Incarnate Word, The Sisters of Charity of the Incarnate Word, accessed August 18, 2022, https://www.ccvichapel.org/about.

[15] Greg Jefferson, "Main Plaza proposal hits roadblock," *San Antonio Express-News,* January 19, 2006, 1B.

[16] Michael Cary, "Main Plaza redux," *San Antonio Current,* February 1-7, 2006, 6.

[17] Jefferson, "Main Plaza proposal hits roadblock," 1B.

[18] "NSA/CSS Locations," National Security Agency/Central Security Service, accessed August 18, 2022, https://www.nsa.gov/About/Locations/.

[19] L.A. Lorek, "NSA plan for S.A. is on hold," *San Antonio Express-News,* January 29, 2007, https://web.archive.org/web/20070929111635/http://www.mysanantonio.com/news/metro/stories/MYSA093006.01A.NSA.32c1dd8.html.

[20] Bradford Shwedo, "Air Force begins development of cyber command," *San Antonio Business Journal,* May 18, 2010, https://www.bizjournals.com/sanantonio/stories/2010/05/17/daily13.html.

[21] "Senator Frank L. Madla Building," Texas A&M University-San Antonio, accessed August 31, 2022, https://www.tamusa.edu/tamusacampuses/maincampus/madla.html.

Chapter Ten

[1] Dan R. Goddard, "Shaping Minds, Building Futures," *Commercial Real Estate Journal,* supplement to the *San Antonio Business Journal,* 2nd quarter 2013, 2.

[2] Goddard, "Shaping Minds, Building Futures," 4.

[3] "Patriots' Casa," Texas A&M University-San Antonio, accessed August 23, 2023, https://www.tamusa.edu/tamusacampuses/maincampus/patriotscasa.html.

[4] "A Brief History of Temple Beth-El," Temple Beth-El, accessed August 25, 2022, https://www.beth-elsa.org/who-we-are/.

[5] SBG San Antonio Staff Reports, "Revisiting San Antonio Hail Storm of 2016 – Second costliest in Texas history," *News 4 San Antonio,* April 12, 2022, https://news4sanantonio.com/news/local/san-antonio-hail-storm-of-2016-second-costliest-in-texas-history; Chrystine Elle Hanus, "Higher Roofing," Features, *Professional Roofing,* April 2020, 43, https://www.professionalroofing.net/Articles/Higher-roofing--04-01-2020/4660.

[6] Hanus, "Higher Roofing," 43.

[7] Ibid, 44.

[8] Ibid, 44.

[9] Ibid, 45.

[10] Steven H. Miller, "Reconnecting Nature," *Constructor,* November/December 2021, 22.

[11] "Fund, build Hardberger land bridge," *San Antonio Express-News,* May 28, 2016, A20.

[12] Nelson M. Wolff, *The Mayor and the Judge: The Inside Story of the War Against Covid,* (San Antonio, Texas: Elm Grove Publishing, 2021) 32-33.

[13] Garrett Brnger, "Navistar cuts ribbon on new South Side manufacturing plant," KSAT.com, March 23, 2022, https://www.ksat.com/news/local/2022/03/24/navistar-cuts-ribbon-on-new-south-side-manufacturing-plant/.

[14] Navistar International Corporation, "Navistar Celebrates Grand Opening of its Benchmark San Antonio Manufacturing Plant," March 23, 2022, https://www.new.snavistar.com/2022-03-23-Navistar-Celebrates-Grand-Opening-of-its-Benchmark-San-Antonanufacturing-Plant.

[15] Brnger, "Navistar cuts ribbon on new South Side manufacturing plant."

[16] W. Scott Bailey, "Fueling change: How Port SA plan could spur more widespread

redevelopment," *San Antonio Business Journal,* June 17, 2022, https://www.bizjournals.com/sanantonio/news/2022/06/17/port-sa-remake-could-ignite-landmark-redevelopment.html.

Chapter Eleven

[1] W. Scott Bailey, "Bexar County to negotiate potential Tower Life Building redevelopment," *San Antonio Business Journal,* August 9, 2022, https://www.bizjournals.com/sanantonio/news/2022/08/09/bexar-county-tower-life-building.html.

[2] Madison Iszler," Developer eyes future for historic buildings," *San Antonio Express-News,* August 8, 2022.

[3] Brian Kirkpatrick, "Residential high rise will offer market value units, join San Antonio's tallest buildings," *Texas Public Radio,* May 4, 2022, https://www.tpr.org/san-antonio/2022-05-04/residential-high-rise-will-offer-market-value-units-join-san-antonios-tallest-buildings.

[4] Ramzi Abou Ghalioum, "$80M Friedrich Lofts development to begin in September," *San Antonio Business Journal,* August 22, 2022, https://www.bizjournals.com/sanantonio/news/2022/08/22/80m-redevelopment-friedrich-lofts-fall-demolition.html.

[5] Madison Iszler, "Creating a village: Gervin-Hawkins' long-sought development on San Antonio's East Side breaks ground," *San Antonio Express-News,* June 29, 2021, https://www.expressnews.com/business/article/Creating-a-village-Gervin-Hawkins-long-sought-16282691.php.

[6] Mitchell Parton, "$130M-plus development envisioned as 'Pearl on the East Side' finally kicks off," *San Antonio Business Journal,* June 28, 2021, https://www.bizjournals.com/sanantonio/news/2021/06/28/long-stalled-development-finally-kicks-off.html.

[7] Madison Iszler, "Wave of redevelopment," *San Antonio Express-News,* August 25, 2022, B1.

[8] Ibid, B6.

[9] Ibid, B6.

[10] "Visitor Center and Museum," The Alamo, accessed August 28, 2022, https://www.thealamo.org/support/alamo-visitor-center-museum.

[11] Nicole Cobler, "Phil Collins' star rises over the Alamo," *San Antonio Express-News,* March 11, 2015, https://www.expressnews.com/news/politics/texas_legislature/article/Phil-Collins-star-rises-over-the-Alamo-6128672.php.

INDEX

Authors

Doug McMurry holds a degree in government from The University of Texas at Austin and began his professional career working in political campaigns and serving in city, county, and state government. He worked in positions of progressing responsibility with not-for-profit construction trade associations before acting as president of the San Antonio Chapter of the Associated General Contractors for twenty-eight years, during which he wrote about construction companies and the growth of San Antonio. A strong advocate of community service, Doug has also led several city boards and committees, and was named an honorary member of the San Antonio Chapter of the American Institute of Architects in 2022.

Michele McMurry, founder of McMurry Communications, holds a Bachelor of Arts in English from Texas A&M University. Her experience as a professional writer and editor spans more than twenty years and includes crafting marketing and public relations content for businesses as well as San Antonio's leading creative agencies. She has also written for and edited trade, food, lifestyle, business, and construction publications in Texas and Florida.

CPSIA information can be obtained
at www.ICGtesting.com
Printed in the USA
LVHW071508120723
751942LV00073B/177/J

9 781958 407035